Off the Radar

Off the Radar

... a man, a plan and a paddock ...

Te Radar

HarperCollinsPublishers

National Library of New Zealand Cataloguing-in-Publication Data
Te Radar.
Off the radar : a man, a plan and a paddock / Te Radar.
ISBN 978-1-86950-765-7
1. Te Radar. 2. Sustainable living. 3. Environmental responsibility. 4. Environmental protection. I. Title.
363.7—dc 22

First published 2008
HarperCollins*Publishers (New Zealand) Limited*
P.O. Box 1, Shortland Street, Auckland

ISBN 978 1 86950 765 7

Cover design by Gra Murdoch, Aqualuna Design
Cover image by Jane Andrews
Typesetting by Springfield West

Printed by Griffin Press, Australia

70gsm Classic used by HarperCollins*Publishers* is a natural, recyclable product made from wood grown in sustainable forests. The manufacturing processes conform to the environmental regulations in the country of origin, Finland.

Dedication

To my parents, Malcolm and Eileen, who provided me with the skills to live this lifestyle, and for Ruth, without whom this book would not be what it is.

Acknowledgements

To all the people who helped me, who gave of themselves, and tolerated me, and humoured me, including you, fearless reader, I thank you.

Contents

1

I'll have the chicken

Marjorie has kindly offered to lend me her walking stick so that I can prod the chicken away from where it has situated itself in the corner of the cage. Now I have two sources of stress: being defeated by poultry, and the fear that any minute a frail 92-year-old woman may topple over and shatter in front of a mortified nation, and it'll be entirely my fault.

The hen, a tiny white bantam, is proving feisty, retreating to the back of the cage, resolutely avoiding my grasping hands and the flailing walking stick, while staring at me with what could be either contempt or total indifference. It's hard to tell sometimes what dark emotions lie behind the beady eye of the fowl.

There are more chickens in the world than any other bird — some 24 billion of them. Little wonder it's a nightmare when they get the flu. I am dealing with three, and I already wonder if that's more than I can handle.

I'm also beginning to suspect that I should have paid a little more attention at a lunch meeting I attended a few weeks prior, at SPQR, a fine dining establishment on Auckland's Ponsonby Road, in August 2007. However, I was a little distracted at the time by the herb-encrusted corn-fed chicken served on a bed of garlic mash and a third glass of a superbly chilled Astrolabe chardonnay.

Jane Andrews and Melanie Rakena are television producers of some note and the founders of Jam TV, responsible for the popular *Intrepid Journeys* series along with the acclaimed *Off the Rails* and

Ice. Now they are interested in me. I know this because they had offered to pay for the lunch.

As the fourth glass of the Astrolabe arrived, they asked me what I was doing over the course of the next 10 months. 'Very little,' I stated. This was followed shortly thereafter by my announcing that yes, I would love to be the host of a television programme about sustainable living that would require me to live in a field and attempt to survive. I may even have followed up this proclamation with the phrase, 'After all, how difficult can that be?' and followed that up with a request for a tiramisu. And an espresso.

The programme they have wheedled me into doing is to be a study in sustainability, or more precisely a study into whether I can become more sustainable given a paddock, some time and a little gumption.

Now, when it comes to issues of global warning, carbon neutrality and peak oil, I have to confess that I am something of an agnostic. I am one of those types who prefers to take their own reusable shopping bag to the supermarket, so much so that I now own several of these bags. But as I often fail to remember to put the bags back in the car after taking them into the house, they reside mostly in a drawer in my kitchen. Thus, not only am I using more shopping bags than I should as I never have the reusable bags with me, I am also now hoarding supposedly better bags.

This show may be a challenge.

'Sustainability' is very much on the tips of people's tongues at the present time, or at least it is when they are not blathering on about real estate. In an era when people are terrified that their use of too many supermarket shopping bags will fuel a rise in global temperatures and threaten their beachside holiday homes, 'sustainability' has been mooted as the little bit you can do as an individual to help save the entire planet from global catastrophe.

I'm all for doing a little bit to help. Especially if it gets me out of doing a lot.

But back to the lunch. It wasn't until much later (well, the fifth glass of Astrolabe) that I might have begun to question what I had let myself in for.

Jane and Mel laid the concept out in greater detail. I would be required to move to a bare paddock in the countryside, with limited resources, including a small set-up fee and a $20 weekly stipend to buy essentials such as any food I couldn't grow. I would then have to attempt to become self-sustaining by hunting and growing my own food over the course of 10 months.

Of course, they explained, you would be free to leave to complete other work that you already have scheduled. This was fortunate, as I was in the middle of completing another television series, called *Homegrown*, about produce and the way it has shaped us as a nation, which in many ways was a fitting parallel to the show they were offering. But, they said, if you aren't away for other work, then it's life on the farm for you.

I even had to write and sign a contract: 'I, Te Radar, will attempt to live as sustainably as possible off this bounteous land.'

They explained that self-sustainability was different to self-sufficiency. 'Being self-sufficient means taking care of everything yourself,' said Mel.

'Whereas being self-sustainable means looking at how you can continue to exist while having as little impact on the planet as possible, with what impact you do have being ultimately beneficial for the future,' Jane added.

Gosh.

During the course of the 13-part series we would also look at the wider issues of sustainability. I was particularly looking forward to having to raise and hunt my own food.

I have long been concerned about where my food comes from and what's in it. As I'd perused the menu earlier at the Ponsonby eatery and contemplated ordering the chicken, I did wonder where it had come from. At that time, there was a stoush between a discount butcher chain and a supermarket giant over who had the cheapest chicken.

Preposterous.

Being able to produce the cheapest chicken is not something, I believe, that could be regarded as a source of pride. However, perhaps my attitude was a little elitist, especially as I wasn't paying for lunch that day.

So it was that I found myself, on my birthday, which also happens to be the first day of spring, filming what would constitute the beginning of the series.

One of the odd things about television is that you are forced to think a lot about the future — or rather, to make wild predictions on camera. This is to provide material for the director to utilise later to deride your naive Utopian dreams, as those dreams become dysfunctional nightmares.

I'm grilled on what I think may happen (I'll grow something), what the problems may be (what I grow will die), and what I hope to gain (hopefully not any weight).

I'd grown up on a farm, and I am keen to return to the land as a small runholder. There is an immense amount of pleasure to be had from toiling on the land.

Having been employed for the last several years to essentially do little other than talk, to suddenly find myself in a position where I can look back at the end of the day and see a physical manifestation of my labours is going to be very fulfilling.

I mention on camera that I don't really have a problem with early mornings. It is this comment that will cause my mother to convulse in spontaneous laughter at a screening some months hence.

It's late September, just a few short weeks after that fateful lunch, and I find myself bidding farewell to my comfortable house in Henderson, West Auckland, and heading to the wilds of rural New Zealand.

Well, Kaukapakapa anyway, northwest of Helensville. It isn't really the wilds of rural New Zealand, as it is only an easy 45-minute drive from my house.

I had only really been to Helensville twice prior to this, as it had always seemed so far away. It's a small town snuggled into the nook of a river, at the southernmost end of the Kaipara Harbour. On one visit I had bought a toilet-roll holder shaped like a dragon. The toilet roll is held between the dragon's outstretched paws. So far I am the only person I have met who finds it adorable.

I had not seen the farm prior to arriving, so when I finally locate it on a bright spring afternoon, there is a sense of healthy optimism in my step as I walk down the long driveway. The farm, my new home, Te Whenua o Te Radar, is more beautiful than I could have imagined.

Jane has excelled herself in finding it. Nestled on a north-facing hill, it runs down a slope to a creek at the bottom, and is sheltered by trees. There's ample barn space, cattle yards and piles of discarded detritus. There's a ramshackle house, which I will not be living in, but which will make great quarters for the crew and editing suites.

At some stage there appears to have been a well-cultivated garden, which is now mostly overgrown, although there are still a few silverbeet plants protruding above the weeds. Feijoa bushes line the back of the orchard. There are lemons, limes and ripe grapefruit. They remind me of boarding school, where grapefruit was served in great abundance to us for breakfast, sprinkled liberally with sugar and devoured. I shan't get scurvy.

If we had been constructing a set for the show from scratch, we could not have created a better environment. The farm is ideal.

The previous owners have left a vast treasure trove of scrap. The farm has clearly been used to run horses, and the remnants of this era still litter the place. Lying scattered around the outside of the barn are piles of bounty that include various lengths of alkathene pipe, coils of wire, sheets of tin and corrugated iron, spouting, piles of battens and posts, lengths of wood in various states of decay, rusting 44-gallon drums, two large stacks of concrete fencing posts, and a magpie in a pine tree.

Elsewhere there are abandoned dog kennels, and piles of firewood lie under a large fibreglass canopy.

Inside the barn it is all a little bit too much like a land-locked, stable-based *Marie Celeste*. On one wall are pinned old ribbons and rosettes, celebrating victories and placings at various horse events. No doubt the prizes were once a source of joy for a child, and pride for a parent. Now they are dusty and oddly touching relics.

Below them is a line of nails that must have once held the harnesses of various horses. The names of these steeds are written on pieces of cardboard stapled above the nails. Whatever happened to Trixie and Bella and Lucky? Did they retire to a green field to live out their days? While I hope that was the case, a little voice keeps whispering 'glue'. Admittedly that is my voice, but it is quickly silenced by a withering look from Jane.

In the next room are four non-functioning freezers. Nearby, towering kitset shelves, constructed from bolted-together steel struts that resemble pieces from an adult-sized Meccano set, slump against the wall. In the corner lie parts of a discarded potbelly stove. I am sure that both of these items will form integral parts of my living arrangements in the days to come. It will be my job to recycle a life from these leftovers of a former existence.

Jane has invested in a farm vehicle, a long-wheelbase Land Rover, which, while roadworthy, has clearly seen better days, most of them some time ago. The Land Rover suffers from what would be described by an expert as 'a significant amount of play in the steering'. In reality this means that keeping the lumbering vehicle on the road, let alone travelling in a straight line, requires swinging the wheel rhythmically backwards and forwards through an arc of about 180 degrees to stop the wheels from scooting off in their own direction. Some vehicles virtually drive themselves, but others need to be driven. The Land Rover is most assuredly the latter.

The indicators are activated by a firm steel switch, which makes a reassuring click when turned on, but isn't technologically advanced enough to turn itself off after cornering. It reminds me of the switches I saw on one of the capsules fired into space during the '60s that was preserved in the Smithsonian Institution. It was a tiny cocoon that contained nothing but hundreds of rows of simple on/off switches. It was gloriously mechanical when considered from the point of view of the high-tech age in which we live.

The Land Rover is the same. It has a GPS system, which in this case stands for Get Past Sixty, that is activated by finding a long downhill slope. Worse, in a fit of practicality triumphing over common sense, the spare tyre is bolted to the bonnet, meaning that whatever field of vision there may have been is now obscured. The only consolation is that if I am to hit someone, then at least I am in a vehicle solid enough to protect me. Maybe they intended the tyre to be a rudimentary airbag?

Almost as soon as I set foot on the farm, I am off again, guiding the Land Rover deep into the heart of suburban Devonport, peering over the tyre, while double-clutching through the gears, and swinging the steering wheel rhythmically through its broad arc.

I have been offered some chickens from somewhere in the middle of this neighbourhood. Thus I find myself at Marjorie's. She has adopted them, or perhaps they have adopted her. No one knows quite where they came from, but they seem relatively tame. They are mine under the proviso that I don't eat them.

I have never dealt with live poultry before, and having to wrangle them out of the cage and into the boxes I have brought with me is a little daunting. The first lesson I learn is that one cannot be timid with chickens. While I know they have to be grabbed — and grabbed firmly — I don't want to break them. Nor do I wish for them to escape and instigate a Benny Hill-esque romp as I chase the flapping birds around Marjorie's section.

I ask Marjorie how I should approach them.

'You have to be very cunning. And don't let them get away — they can run like billy-o,' she says.

I'm expecting a fluttering flailing of wings and claws, but all I have to do is reach in and gently lift the first of the chickens up. Black, but with ginger feathers around its face and neck, it looks like a little pheasant. It stares at me as Marjorie says, 'They're very lovely, friendly birds.'

'Yes, I can see that. They're about the friendliest bird I've ever encountered, and I've encountered a few friendly birds in my time, Marjorie,' I joke, as I put the chicken in the box that I have brought along.

'Yes,' she says.

The procedure is the same with the second hen, which is large and red and totally unfazed by being manhandled. This is all going a little too easily I think.

Unsurprisingly, things become more difficult at this point, and I find myself hoping not to hurt Marjorie as I try to wrangle another bird. After what seems like an eternity, I'm clutching the chicken to my breast and handing Marjorie back her stick.

With the chickens collected I am the proud possessor of poultry. Already I feel more rural, and more sustainable. I have chickens, and chickens mean eggs, which clearly these chickens are producing in abundance, as Marjorie kindly offers me several that have been laid over the preceding days.

Boxed and deposited in the back of the truck, it's off to the farm to install them in my egg-production facility.

On the way back I am almost delirious with euphoria. The reality of my reality is sinking in. That I am to front my own primetime television show is exciting enough, but to be given the opportunity to live off the land — to hunt and farm and grow my own crops — is a privilege that not many are offered.

And perhaps the best thing about the programme's concept is that while I am being challenged to survive in the harsh environment of a meadow, I am to do so alone. Therefore, no matter how bad at it I am, no one can vote me off.

With night closing in, it's time to wrestle a living space from the pasture and construct a habitat. It seems a wonderful parody of the pioneers of old, who came here and had to hack a living space from the rugged bush. No gentle fields of England lay here awaiting them.

Whereas they had to use axe and fire to clear the land, I merely have to stroll to what seems like a suitable spot in a paddock, stomp around in the knee-length grass to establish I haven't chosen a hollow, and begin to denude the area in preparation for raising a shelter.

To my surprise I manage to get the weed-whacker going, and in a cloud of smoke (another homage to the scrub-burning pioneers) I mow an area that I assume is big enough to erect the tent.

I have never erected this tent by myself before. Purchased by my father at the end of the 1984 Feilding Scout Jamboree for the princely sum of $150, the tent is older than I am. Ex-army, made of canvas, with deceptively heavy poles, it is not the kind of tent one takes hiking. In fact, the entire thing, poles included, weighs in at over 50 kilograms.

I struggle with the tent. It's made more difficult by the fact that they are filming me erect it. My efforts are finally rewarded, and the proud khaki galleon sails upon the grassy sea.

Stepping out of the tent, I experience a moment of pure euphoria. It is quelled only by Jane saying, 'We'll have to do it again.' They forgot to press record.

Later, lying in my sleeping bag, in the tent, in the paddock that is now my home, I realise there's no such thing as a free lunch.

2

On the snail trail

I am standing in the middle of the garden, holding a snail delicately between my fingers, and experiencing something of a conundrum. I consider the snail a pest, whose desire to eat the plants that I rely on to survive means that it is worthy only of eradication. However, it is also considered by many to be an excellent source of food, and if there is something I am short of, it is food. What am I to do?

While it may be possible for a man to live on a diet based entirely on eggs, silverbeet, potatoes and grapefruit, I am not that man. Even though I certainly won't develop scurvy, and have to suffer the ignominy of losing my teeth, diarrhoea, and bleeding from all my mucous membranes, that diet would cause me to wish I did, if only to relieve the monotony.

Many of the vegetables I plan to plant won't be ready for up to 90 days. Leeks take a staggering 21 weeks. That's five months. Who can wait that long? I don't think at this juncture that I'm that desperate for a leek.

I've been living in a world of seductive convenience. When it comes to buying fruit and vegetables I don't worry about the vagaries of seasonality, or weather, or pests, or the length of time a plant takes to mature. I go to the supermarket or the greengrocers, and whatever I want is generally available.

What I need now, though, is an emergency garden, one that will bear fruit — or vegetables actually — as soon as genetically possible to tide me over until the rest of my crops have matured.

I have never really grown anything before, bar a successful crop of cherry tomatoes the previous summer. I have no idea where to begin, or even what is possible.

Fortunately, a short drive north of the nearby village of Matakana, up in the low hills, nestles Rainbow Valley Farm, the El Dorado of all things sustainable.

The land the farm is situated on was once deemed by local farmers to be nothing more than 'rubbish land' — good for little more than growing the weeds and pests that proliferated there. Then, 20 years ago, up the long winding valley a house-truck hove into view, ground its way up the hill and down the steep, winding driveway, and out onto the heavy clay soil that was a veritable bog in winter and like concrete in summer leapt Joe Polaischer and Trish Allen. They were the new owners of this 20-hectare block and they had arrived with the goal of becoming self-sufficient.

Now, descending down the driveway, it really does seem as if I am entering an abundant oasis.

Crossing the lawn, next to a lush vegetable patch, I approach the house that Joe built. Tucked into the side of a hill, the long, low-slung house features thick adobe walls, and a roof upon which a luxuriant crop of grass grows. I'd often seen it on television and in magazines, but in the warm spring sunshine it is even more impressive than I had imagined.

Joe hobbles past a few idle chickens to meet me, apologising that only a few days ago he had a hip replaced, and isn't his usual self. I would never have known, as he looks pretty damned sprightly. In fact, both he and Trish positively glow. There is a light that shines in their eyes and in their cheeks. Smiles seem to fall from their faces. It's little wonder they and the farm are spoken of in reverential tones by those who know them.

After a cup of tea and some cake, Trish offers to take me on a leisurely perambulation of the grounds. 'We've planted a few trees since we've been here,' she says as we stroll through the orchard. They certainly had. Where once there were weed-covered slopes, there are now over 13,000 trees, including over 800 fruit trees. 'What we've created is an edible landscape,' she says with a certain degree of pride.

Everywhere, something useful grows. Through the orchards all manner of birds strut and preen. Large gardens overflow with vegetables of many hues.

Rounding a corner we are confronted with a huge, richly aromatic, sacking-covered mound that sits squatly, steaming. 'I think that's the best compost pile I've ever seen, Trish,' I proclaim.

'That'll get up to sixty degrees at its peak,' says Trish. 'The key to great compost is layering the carbon and the nitrogen. Nitrogen is your green material — chicken manure, coffee grounds — whereas the carbon is your straw and twigs, that kind of thing.'

I had no idea there was a science to composting. Why are we not taught these things? Perhaps I had been and had forgotten. Until this point in my life there hasn't been a great call for compost knowledge. I don't think I have even had a conversation about it prior to this. I am beginning to suspect that I might be slightly unprepared for the next few months.

Trish explains that Rainbow Valley Farm was designed in line with their belief in permaculture, so that everything on the farm has a place and a function.

Permaculture, a contraction of 'permanent' and 'agriculture', is a system that was initially formulated by a couple of Australians in the 1970s. In its most simple form, it is the belief that one should allow nature, and the natural cycle, to provide solutions to problems. By integrating various aspects of nature, you should be able to eliminate a lot of the work normally considered essential in farming, or gardening, thereby saving energy and eliminating waste.

Their toilet epitomises this. Over the years I have experienced many toilets that have taken my breath away. Literally. I have always been a little dubious of any toilet that won't flush. Piles of human effluent sitting in tanks or holes not far beneath me, producing odours and insects — or worse, being visible — has never appealed. Mercifully, long gone are the long drops of old. Not only are they illegal in many places, but there is no way they can compare with the modern composting toilet, the likes of which is now being explained to me by Trish.

They are breathtaking in the most pleasant way. Large, airy and colourfully tiled, they are as far from the image of what a composting toilet would look and smell like as I can imagine. They are the ultimate in recycling, as the composted manure returns to the soil from whence it originally came, where it fertilises new crops, and so the loop begins again. I would like to ask if I can try it out, but I am too embarrassed to do so.

As I stand inside their main vegetable garden, surrounded by a little picket fence, eating snippets of plants that Trish has plucked from the beds, I ask: 'How long will it take me to get something like this garden here happening?'

'Oh, not long at all,' replies Trish with a sense of unbridled optimism that indicates she clearly doesn't know just how little experience I've had with plants.

I have often tried to establish gardens in the past, with limited success. More often than not, after an impulse stop at the nursery, I have ended up with trays of seedlings that I've never quite got around to planting, and I then spend the next few weeks watching them slowly die in their little pottles.

Trish hands me the orange flower of a nasturtium to sample. It tastes slightly floral. She smiles at me again. I'm smitten in the most wholesome of ways.

As we talk, Joe, considered to be one of the world's great permaculture teachers, bustles about, instructing someone here, planting something there. People from all over the globe have flocked here to be taught about life and living, and I can well understand why. Trish and Joe's optimism and generosity are infectious, and I leave determined to create my own Eden.

All I need are some plants.

While beginning my garden entirely from seeds may be cheaper, what I need are seedlings. Thankfully there is yet another handy woman who can help.

A quick trip up the road in the Land Rover brings me to a jaunty yellow sign on a sharp bend just south of Kaukapakapa. It marks the entrance to Weathersfield Organics, a vegetable and herb seedling-raising enterprise developed and run by the enchanting Pru Broughton.

'You see, Radar,' Pru says, as we stand squirting water in the general direction of tables of lush green lettuce seedlings, 'conventionally bred plants are bred for their eye appeal in the supermarket. It's all about size and colour, and their ability to last in cool stores. They're not bred for their taste appeal, so you don't necessarily get a lot of flavour out of them.'

Having pottered around with plants her whole life, Pru and her husband Roly were in the process of setting up the nursery in 2002 when Roly died suddenly. Determined to carry on their dream she persevered. From a few tables covered with shade cloth she has built Weathersfield up to the point that it now has several shade houses and contracts for virtually as many seedlings as she can supply.

As demand grew, Pru decided that shipping the seedlings she raised in Kaukapakapa all over the country would not exactly be in line with her organic principles and beliefs in local produce. The answer was to franchise the operation.

This adherence to a philosophy of sustainability is something I will encounter repeatedly over the next few months.

'The most important thing about an organic garden is to keep feeding the soil, and the soil will do the rest,' she says. 'I like to think of it this way: that the soil is the supermarket, the roots do the shopping, and the leaves do the cooking. If there's nothing in the supermarket, then there's nothing for the shopping basket, and nothing to cook, so the plant's not going to grow. You have to feed the soil, preferably with high-quality organic material, not chemicals.'

This belief was for a long time one of the central areas of mockery directed at the organic types. Why, when we have science providing us with supposedly quicker and easier ways of improving the soils and protecting crops, is any of this necessary?

The main problem is that, other than any questions as to their negative environmental impact, these fertilisers and the pesticides are manufactured with, and transported by, cheap fossil fuels. With the fear that the era of cheap fossil fuels may be over, those who spent years being regarded as flakes because of their emphasis on natural, organic methods don't seem so flaky now.

As I am leaving, Pru wishes me luck, and says that gardening's pretty simple, and that even a child could do it: 'My grandkids graze straight from the garden. They graze on the spring onions, they graze on the parsley, and they graze on the silverbeet,' Pru informs me. 'Getting down and dirty is half the fun of being a kid.'

'It's half the fun of being an adult,' I respond.

As we laugh and load the Land Rover with trays of seedlings, I make a mental note that not only do I have to worry about slugs and snails and caterpillars and chickens, but I'm now going to have to go to all the effort of building a childproof fence, as the last thing I want is to have my crops raided by a voracious horde of grazing children.

Actually, that isn't quite the last thing I want. The last thing I want is for all my crops to die, although now that I have them back at the farm and planted, if they do die, at least it will be in their beds.

With everything planted, I stand back and survey my handiwork. Two raised beds of seedlings flutter gently in the wind — one, a plot of organic plants; the other a group of ordinary, conventionally bred ones.

Faced with the question of whether I will embark down the path to total organic production in my season of sustainability, I have decided instead that I will hedge my bets and trial both organic and conventional methods. This way, I can judge which seems best, and reduce the risk of one crop failing by having two crops.

All I need to do now is to protect them from the myriad of dangers that will confront them, including predators such as slugs and snails and birds and caterpillars (not forgetting grazing children).

Having been told that neither the slug nor the snail has the ability to swim, I have scrounged a few lengths of old spouting, from which I plan to create a simple moat filled with water around my vulnerable plants.

I will never know if the theory behind the moat system is correct, as regrettably I never manage to implement it.

I assemble most of the parts required, and I even get as far as to lay those I have in place. But then I stop. Well, not stop as such, but just postpone finishing it.

If there is one thing that is going to cause the failure of my efforts to live off the land, it isn't going to be starvation. It's the fact that I am an inveterate starter of projects that never get finished.

I'm a hopeless procrastinator.

I often wonder whether or not procrastination is a mental illness. It would be grand if it were, for at least then I'd be able to confidently state: 'I'm not lazy, I'm sick.'

On several occasions I have even found myself procrastinating by sitting at the computer reading online articles about how to stop procrastinating. Most of them seem to be written from the intolerable viewpoint of the non-sufferer. I guess if the real procrastinators were writing them, they'd never get written, unless the procrastinator was writing an article on procrastinating as a form of procrastination in and of itself.

So with no moat, I have to consider what to do about the snails. Lifting a snail in my hand closer to my face, I inspect it thoroughly. I realise that in the interests of the sustainability experiment, I should at least consider harvesting and eating them. All I need to do is collect the snails, feed them on some flour, or a little carrot, for a short period to 'clean them out', and then fry them in a little garlic and butter, and voila! I will have a nourishing and, I suspect, a healthy meal. I can't imagine that there is too much fat on a snail.

There are several reasons why I do not do this.

Firstly, I have no idea how to kill them.

Secondly, not even the chickens seem interested in eating them, which I take to be a sign.

Thirdly, they are snails.

And even if I do eat them, I will still be faced with the problem of what to do with the slugs. I have never heard anyone utter the phrase, 'I feel like a good feed of slug.'

While I don't mind a snail, I cannot abide a slug. I guess some of it has to do with the fact that you can avoid touching the body of the snail as its shell provides a handy handle. A slug is unadulterated mucus.

I guess they couldn't be any worse than the misleadingly named Chinese delicacy, the sea cucumber. While it lives in the sea, the sea cucumber is most assuredly not a cucumber. It's an animal, a fact of which I am aware as I once boldly pointed to it on a menu, and said 'One number 174 please. Yes, the casseroled sea cucumber.'

What I didn't know at the time was that the creature breathes by drawing water in through its anus, extracting the oxygen from it, then expelling it from the same orifice. Had I known this prior to ordering it, I may well have asked, 'Do you have any snails?'

Eventually I decide that the easiest way to deal with both the snails and the slugs is to use something that is the answer to so many of life's other little problems — beer.

Poured into a container that is then placed in or near the garden, the beer entices the snails into the container, whereupon they fall into the beer and drown. What a way to go.

While the beer seems to work for a while, its main shortcoming is my desire to consume more of the beer than the snails do. The thought of pouring hard-earned beer into these vessels in order to drown snails seems like a waste of good beer to me, for if there is anything that needs drowning, it's my sorrows.

3

Three fat turkeys

The phrase 'Apparently you are going to let me shoot your gun' is considered by some to be at best a little effete, and at worst, decidedly camp. Nevertheless, it is my opening gambit upon meeting the chap who has been conscripted to help me gain a little game.

Shawn is a nuggety jack of all trades who happens to live just up the road. Goodness only knows what he thinks as he arrives in his ute to meet the television guy who wants to do a bit of hunting. Whatever it is, I cannot imagine my greeting will inspire him with any degree of confidence in my abilities.

Despite this, he agrees to take me out into the field in pursuit of dinner.

'There's good eating on a turkey,' he promises.

'Really?'

'Oh yeah, and we'll find them easily, no problem at all.'

The only problem is that he doesn't inform the turkeys, so as we drive around his mate's farm on a warm spring morning, there is nary a sign of our feathered prey.

Hunting with a TV crew in tow lends a certain air of expectation and pressure to the occasion. There is an imperative to find the game, because having a cameraperson, a soundperson and the director amble around, filming me standing around, might be good for my ego, but it is not great value for money. Nor does it create overly riveting television.

After much scouring of farm, combined with some head scratching and the use of a considerable amount of colourful language to describe the turkeys' intelligence, lineage and non-appearance, we finally locate some. The only problem: their location. This particular flock has decided to loiter next to a herd of cows, which in turn are located next to a house. These are not ideal conditions under which to unleash a barrage of shot at the feathered food source.

'No problem,' says Shawn. 'We'll move them.'

'The cows?'

'No, the birds, you wally.'

'Oh. Of course.'

The concept of shooing game to a more convenient locale is not something I have previously experienced, as one generally sneaks up on whatever it is one wants to shoot in order for it not to run or fly away.

Leaving the crew by the ute, Shawn and I stroll in a wide circle around the birds in order to begin shooing them away from the cows, away from the house, and towards an empty paddock.

'They're quite a popular trophy species in the States,' says Shawn, as he eyeballs a bird that seems to want to walk the wrong way. 'Guys'll sit under trees all day waiting to shoot a big male so they can collect his beard as a trophy.'

Turkeys have beards? I had no idea.

I've only ever been close to supermarket turkeys before, and they don't even have feathers, let alone beards.

For a moment, I think Shawn might be pulling my leg, but he explains that collecting the long, coarse, hair-like feathers that grow from the males' chests is akin to trying to shoot a stag with big antlers. The older the bird, the longer the beard, and the more prestige there is for the hunter.

'You know', says Shawn, 'you can just sneak up to them at night when they're roosting on a fence, then just grab them by the legs, chuck them in a sack, and take them home to fatten them up.'

I tell him that to be honest, the thought of wrestling a turkey off a fence in the middle of the night may be just a little too adventurous for me at this juncture, and that I might just stick with shooting them.

'Fair enough,' he replies.

The turkeys prove very obliging, warily staying some distance from us as they scuttle off in the general direction of where we want them to go. It is a peculiar sensation to herd a flock of wild birds that I will soon be merrily blazing away at with a shotgun. Knowing what is in store for some of them, I feel oddly callous. Then again, is it any different from rounding up a mob of cattle to send them off to the meatworks, or to bring in a sheep so that it can be butchered for the freezer?

Although I was brought up with hunting, I was never what could be described as exceptionally proficient at it. I have stalked rabbits, eradicated possums, and attempted to shoot various waterfowl, predominantly to no avail.

The last time I went hunting any kind of bird, it was the Canada goose, a bird that was introduced to our country as a gift from US president Theodore Roosevelt. While he probably thought it would make an excellent game bird, many people now consider them a plague, due to their ravenous appetites.

It is said that one bird can eat as much each day as a large lamb, and what goes in must come out. Their faeces leaves paddocks so soiled that cattle are loathe to eat the grass.

While a few geese may not be too bad, having mobs of hundreds of them descend on paddocks is more than a little problematic.

So one night several years ago, my brother and I set out to thwart the birds from settling on the paddocks at the back of our parents' farm. Squatting in the dark, nestled behind a low line of manuka growing out of a drain, we waited for the distinctive flying V of the birds to pass over us.

In the quiet of the night, their honking could be heard in the distance, gradually growing in intensity as they flapped in over the manuka. Standing, we unleashed a barrage of shots at the flock.

Suddenly my brother shouted, 'Run!', and began to gallop away.

I stood where I was, as I'm often a little indecisive in situations like this.

The cause of his consternation soon became clear, as one of the birds he had shot crashed heavily onto the very spot where he had been standing. It plummeted to the ground with enough force to indicate that had my brother still been standing there, he could well have been seriously injured in a most embarrassing manner.

It probably wouldn't have been as embarrassing as the goose-based injury that was inflicted on man-model Fabio, who was once struck in the face by one of the birds when riding on a rollercoaster, but it would still have been very painful.

As it was, I was the one who was nearly injured — I was laughing so much I very nearly gave myself a hernia.

My brother finally allowed me to go shooting with him again, only it wasn't birds we were after this time, it was fish.

As the Waikato River floods and the water pours over the spillway at Rangiriri, it causes Lake Waikare to flood the rear of our parents' farm. Here the low-lying peat flats are at best only half a metre above lake level, so as the waters rise and the pasture becomes flooded, koi carp swim in to take advantage of the pasture.

My brother and I responded to these trespassers by donning waders, arming ourselves with shotguns, and going fish shooting.

Fish shooting requires a special set of shooting skills, as you not only have to stalk the fish, wading slowly through the murky waters without falling into a hole and scaring them off, but once they are located you need to carefully judge the angle of deflection through the water to actually hit the fish.

Described by environmentalists as the possum of the waterways, these colossal goldfish have reached epidemic numbers in the Waikato, and cause so much damage to waterways and native species that an annual tournament, the New Zealand Bowhunters Society's World Koi Carp Classic, is held to keep numbers down.

In 2006 the hunters bagged over 3500 fish, weighing in at a staggering 8.5 tonnes, in just two days.

The evening my brother and I hunted them, a single blast from a shotgun would cause several to float to the surface. There were so many of them that it was literally like shooting fish in a barrel.

It was just a shame that, unlike the turkeys, they were virtually inedible.

By now, the turkeys are safely away from the cows and about to disappear into a row of scrub. Cradling the shotgun in the crook of my arm, I withdraw two shells from the cartridge belt buckled around my waist.

I have always loved donning cartridge belts. Filled with shells, they have a definite weight to them, and when strapping on the heavy bandolier of birdshot I've always felt a little like some kind of pseudo-revolutionary. Juvenile, I know.

The shells slide into the breech of the twin barrels with a reassuring thunk-thunk. I snap the barrel closed, take a slightly too-wide stance, and bring the gun up so that I am looking along the barrel at the mob of birds, which is about 30 metres away.

As this is the first bit of hunting we are engaged in for the show, I feel a certain amount of pressure not to miss. One doesn't want to be seen on national television missing what many would call a relatively easy target.

Despite this pressure, I aim to shoot the most succulent-looking bird in the head. Due to the tiny size of their heads in relation to their hefty bodies, this is a little ambitious. I've made the decision as much for humane reasons — a clean kill is something a hunter morally owes his prey — as for practical ones. I want to avoid peppering its body with shot, as there is little so dispiriting while eating game as breaking a tooth on a small piece of lead or steel.

With a grim sense of resignation, I thumb the safety catch off, line up the head with the steel nubbin of the sight, caress the trigger with my finger, and squeeze. The gun kicks, the noise deafens, the turkey flaps, the acrid smell of gunpowder fills the air as I swing the barrel around to follow the rapidly departing flock, select another, aim for its head, and squeeze the second trigger.

This time though, something unexpected happens. It appears that I have somehow managed to line two of them up and have shot them both. The only hitch is that as we walk towards them, one is still flapping a little, and I'm not sure that it's entirely dead.

This is not a good thing.

I don't really want to shoot it again, but if it isn't dead, I don't want it to suffer.

'Break its neck,' Shawn says.

'Righto,' I say, before scratching my head and looking confused. While it is no doubt a good idea in principle, I'm not entirely sure how to go about it.

'Pick it up by the neck and give it a good swing around,' Shawn says.

The turkey is a hugely unattractive bird. It has tiny, beady little eyes. And then there is the snood, a fleshy protuberance that hangs from the top of the bill, like a fleshy flaccid finger, and which they can extend or contract seemingly at will. Their long neck culminates in a featherless head that appears to be made of leftover scrotal skin, and it is this clammy-looking skin that I will have to grasp in order to put it out of its misery.

It's warm, and a bit squidgy, and as the bird is surprisingly weighty, I struggle slightly to swing it around. Grasping it firmly with both hands, I whirl it around, feel a cracking somewhere in the neck, and the bird stops flapping.

The whole process has taken less than 30 seconds.

I turn to Jane and say, 'I suspect we won't see that bit on the telly.'

'No', she agrees. 'We most certainly will not.'

By the end of the first day of television hunting, I have bagged three birds for two shots. While I'm chuffed with myself, I'm also a little remorseful for the birds, who hadn't really chosen to give up their lives for my dinner.

Of course, there is a lot to do before I can gobble these gobblers. Shooting them is really only the beginning of what is quite a lengthy process in transforming them from a bird wandering around a paddock to a tasty treat on a dinner plate.

Struggling to carry the weighty birds back to the ute, I wonder why we couldn't have shooed them closer to the vehicle in the first place. As I throw them on the back of the ute, their eyes gleam dimly in the sunlight.

They continue to stare accusingly at me when I get them home, and also as I lug them across the paddock to a handy tree stump a suitable distance away from my tent. The stump will make a useful field table.

I've never plucked anything before, and I am not entirely sure where to start, as the turkey requires a serious amount of feather shedding to strip it of the 3000 plumes protecting its pale white carcass from the elements.

There seems only one way to do it, and that is to grab handfuls of the feathers and start pulling them off. This is easier said than done, so in the end I resort to skinning the birds.

For a short period a few years ago, I largely refrained from the guilty pleasure of eating chicken skin, as medical experts cautioned that it was a major cause of heart disease. A study was then published that said that by abstaining from consuming the skin of the fowl, I might have increased my chances of contracting cancer of the bladder. Is there no justice? It was most confusing.

The same study declared that people who ate bacon five or more times a week were 60 per cent more likely to develop bladder cancer than non-bacon eaters. I suspect this wouldn't unduly worry people who ate that much bacon, as they would most likely die from obesity before the bladder cancer killed them.

I put the article down and made the decision that in the interests of my health I would stop reading articles about researchers who claim that they have discovered new ways to prevent me from catching cancer, heart disease, or fatness.

The reason for this was simple: the stress these articles caused me by constantly contradicting one another exponentially increased my chances of death by either a heart attack or a brain implosion.

The biggest health danger I currently face, however, is slicing myself open while skinning the turkey. With the programme designed to be screened when children are watching, there is very little of the process that we can show. A great deal of animated conversation ensues about the correct level of graphicness that will be allowed. The answer is as little as needed, but enough to show that living off the land isn't all strawberries and apple cider.

I consider this a great shame, as it is a glorious part of the process, but then I've always been fascinated by taking things apart to see what makes them tick, although I'm not such a great putting-it-back-together-afterwardser.

As a child I used to engage in a small amount of amateur dissection of dead animals, predominantly rabbits and cows. I would open them up for a rummage — strictly in the interests of science, of course.

There is such a range of colour inside an animal — the green sheen of the intestines, the dark, almost purpley red of spleens and kidneys, the pink of the lungs. Every organ has its own particular beauty.

In the turkeys' case, the great revelation is their gizzards, where all the things they eat are packed in then ground down with little stones. I can see exactly what the birds have been eating, which seems to be mostly grass seeds, snails and clover. I get the feeling that these birds are going to taste sensational.

After a considerable period of time, I have the birds mostly prepared. At one stage I lose my knife, only to discover I have left it in a turkey.

Once the skin and fur or feathers have been removed, it's always startling how little of an animal is left. The turkeys, which had looked so large, are now a third of their original size, yet there appears to be enough meat on each bird to last several meals. Their thighs, though not as plump as those of a store-bought bird, whose legs fatten with unlimited supplies of food and little to no walking, are lean but meaty.

The real shock is the breasts. Two large, lean fillets cover each bird's chest. With the price of turkey breast meat in the supermarket at well over $20 a kilogram, I am wiping feathers and assorted bits of bird gunk off a small fortune of bird flesh, harvested in a single morning for the princely sum of two shotgun shells and a little time.

The only question that remains is: how will they taste?

On top of a concrete slab in front of my tent I have reconstructed the potbelly stove that was lying in the barn. It proves to be the ideal heat source over which to simmer two meaty thighs in a pot with potatoes, foraged from where they had self-seeded in the compost bin, some rosemary and various other herbs, all mixed in with a good dollop of red wine (after all, a man can't be expected to rough it without a little plonk).

The casserole is delicious. If this is to be the quality of food that I will be enjoying during this sustainability experiment, then things are beginning on a mouth-watering note.

As I sit relaxing in my thinking chair, with the last of the day's warm spring sun shining down on me, I look around at the expensive lifestyle blocks that surround me and wonder how many of the people who live there will be roughing it tonight under canvas, their bellies full of fresh bush turkey.

Only one of them. Me.

As far as lifestyles go, it doesn't get much better than this.

4

The cows come home

When the first cows to arrive on our fair shores were being landed in the Bay of Islands in 1814, one onlooker noted that the excitement of the local Maori, who had obviously never seen cows before, soon turned to fear and confusion as the cows became unruly. The Maori, convinced a supernatural monster had been released upon them to kill them all, fled in terror. I know how they felt, for on a number of occasions I too have been traumatised by cows.

The cow can be a belligerent animal. If they do not wish to do something, it can take much cajoling to change their minds. This fact is foremost in my mind as I find myself standing behind a small Jersey cow, attempting vainly to physically propel her up the ramp of a truck.

Despite my best efforts, the cow in question, a plump, doe-eyed Jersey called Harriet who appears to be of middle age, is happily refusing to move.

This is somewhat annoying as I am paying by the hour for the truck I am attempting to load her into, in order to transport her back to my estate.

She is one of two cows I have purchased from the owners of a neighbouring lifestyle block. The other, a young wild-eyed heifer named Coco, appears to be some kind of Angus cross, and is what could politely be described as frisky. Harriet is Coco's stepmother, and both are supposed to be charmingly docile, and pregnant, which means that in short order I should be provided with both milk and an increased herd.

One of the many problems facing the hobby farmer is what to do with surplus stock which, due to the small size of their herds, they have no doubt formed a bond with. Putting them into their freezer is usually not an option, as the animals are often akin to family pets, and few people want to eat their pet.

Likewise, the thought of condemning their surplus animals to 'the works' is equally abhorrent. Fortunately, their overstocking problem is the solution to my lack of stock.

There is only one hurdle between myself and agrarian perfection. The lifestyle-blockers do not have a loading race, let alone any cattle yards, that are accessible to the truck. With a light drizzle falling, the paddocks are too greasy to risk the very real possibility of a fun-filled and expensive afternoon extricating a stuck stock truck. Thus I find myself attempting to load the cows onto the truck parked on a dry mound near the gate.

However, as is often the case with cows, I have reached that point where what they wish to do, and what I wish them to do, is at an irreconcilable impasse.

So far we have managed to corral them, place halters on them, and lead them across several paddocks to where the truck is parked, but getting them up the ramp is proving impossible.

We try to tempt them with a little hay. While they are willing to nibble the hay when it is dangled enticingly beneath their noses, they aren't interested in allowing it to act as a scrumptious lure to entice them up the ramp.

At this point, the lifestylers reveal a secret weapon: a bucket of cow nuts. The cow nuts look like molasses-coated popcorn, and are a little like cow candy. Judging by their effect though, they seem to be more like the crack cocaine of the bovine world. The cows fight to force their heads into the bucket, their long tongues scooping out the nuts.

As I move the bucket towards the ramp and attempt to seduce them forward, they remain where they are and try to get the nuts by reaching out with their long tongues.

Being licked by a cow is one of the most agreeable experiences one can have with a bovine. They have deliciously rough tongues, so the sensation is akin to being licked by warm, fleshy sandpaper. If you have never been licked by a cow, I strongly suggest that you sally forth and be licked forthwith. If you don't have your own cow, it's best to ask the owner first. Believe me — it's easy for these things to be misinterpreted.

The nuts do their trick, and within moments I am able to entice Harriet up the ramp and into the truck. Coco, though, will have none of it, and steadfastedly refuses to even consider heading up the ramp.

At one stage she appears to change her mind (it is, after all, a woman's prerogative), so, thinking I could administer a little incentive to hustle her along, I deliver a sharp slap to her hairy flank. This certainly causes a little hustling, albeit of the reversing variety, followed by a spirited attempt to bolt from the scene, which is halted only by the firm hand of the truck driver, who manages to hold on to the rope to which she is tethered as she dances a merry high-hoofed jig.

My feelings of abject stupidity are heightened by the fact that I know that, despite my protestations, this humiliating bovine-based failure will make it onto the telly and be witnessed by the nation. This is only slightly less demeaning than the knowledge that it will also be seen by my mother, the woman who has taught me nearly everything I know about cows. She will find my inability to get a cow into a truck a source of great and ongoing amusement.

After much frustration, Hamish the cameraman, Frank the soundman, and Jane the director all abandon their posts — leaving the camera rolling, of course — and together we all heave Coco from behind, as the driver hauls her from the front with her halter. Begrudgingly, she totters up the ramp and into the truck, where she stands looking at us as if to ask what all the fuss was about.

I love cows. I was brought up with hundreds of them on a dairy farm, and as much as they were productive economic units, they were also our friends. They have such wonderful and diverse personalities. In a herd of cows there will always be those who stand out.

Like any group of women they have a rigid hierarchy and social structure. There are the bullies, and there are the bullied. There are ones who are friendly and ones who are totally disinterested in you. Generally, I've found the latter are the more numerous ones, both in cows and humans.

As a child, bringing the cows in for milking meant that I had a great chance to observe their foibles. There would always be several who would distinguish themselves from the mass of the herd with odd little personality quirks.

Penny had a penchant for eating your pen if you should be in the paddock using one; number 214 would walk down the steps into the milking pit where you stood, and wait to be patted; Rosebud, a bright ginger minx, simply loved to be loved. When she died of a mysterious illness, my father was so upset he got drunk. It was the only time I recall him drunk when I was a child. Such is the force of their personalities.

And now I have my own herd. Happy days are here.

With Harriet and Coco safely stowed, all that is left is to wave a cheery goodbye to the lifestylers and return home.

Naturally, having been reluctant to board the truck, the cows are now reluctant to leave it. I need to convince them that the grass is in fact greener outside the truck, and that as the cost of the truck is continuing to mount, now is a good time for them to disembark.

They finally poke their noses out, survey their new domain, and stagger down the steep ramp and into a lush pasture that has not been grazed for some time. Judging by the fact that they immediately begin eating, they are clearly unmoved by their recent move.

Off they wander, grazing as they go, safe and sound.

As the rain continues to fall, the man with the truck and the crew all depart, and I am left alone with my cows. Eventually, they too decide that they will leave me alone, and clearly keen on exploring their new environs, they wander down the hill, while I hole up in my tent.

As the rain eases, I think it prudent to check on my new herd. I shouldn't have.

Harriet and Coco are now standing at the bottom of the hill, next to a line of trees that had previously been concealing a creek. The trees are no longer concealing the creek, as it has risen dramatically and is now a swift torrent of brown water flowing across a wide swath of the bottom of the paddock, deep enough almost to submerge the top of the fence posts.

The cows have wandered onto a small mound that is now completely surrounded by the floodwaters, and they are marooned. This wouldn't be so bad if it wasn't for the fact that the creek appears to be continuing to rise.

Brilliant. I've had possession of my herd for only a few short hours, and it appears they are to be drowned.

That the camera crew is not here to record my misfortune serves only to compound my misery, because as tragic as the event would be, it would no doubt make for wonderfully dramatic television.

Much to my consternation, Coco seems intent on plunging into the swirling waters, which will most likely sweep her into the fence where she will become entangled and probably drown. There's little I can do except suggest to her in a reasoned tone that this might not be the best idea.

I am certainly not going to plunge in and attempt to save her, as having a cow drown is certainly less humiliating than being drowned trying to save a cow. No doubt I could cling to her and she might carry me to safety, but I'm not sure Coco will be as obliging as cow number 569, the bovine hero of the great Manawatu flood of 2004.

As the flood washed across the district, Kim Riley, a local farmer, was caught in a similar predicament to mine, except that she found herself enduring it in the early hours of a very dark morning. Attempting to move her 350 cows to safer ground, she and the cows were caught in the turbulent floodwaters and swept away. The swollen river twice submerged Kim, and unable to swim to safety, she kicked off her gumboots and struggled to remain afloat as she was swept along.

Suddenly cow 569, which Kim described as 'an ugly old thing', hove into view, allowed Kim to clutch onto her neck, and together they struggled to safety. The feat made headlines around the world, and Kim went on to write an internationally bestselling children's book about the incident, entitled simply *Cow Power*. There was even a sequel, entitled *Baby Cow Power*, about the calf that cow 569 bore several months later.

I suspect that, if I try an audacious rescue attempt for my cows, all I will end up with will be a small footnote in the obituary section of the newspaper, or worse, in the quirky news section. If I am very lucky, I may become a trivia question in a pub quiz somewhere that asks: 'What animal was comedian Te Radar trying to save when he died?'

All I can do is return to the tent and hope for the best.

The following morning, the waters have subsided, and Harriet and Coco are standing around, chewing their cud, as if nothing had ever happened.

5

Itsy-bitsy spider

I am examining and selecting pieces of timber from the various piles that are lying around, and am aware that I may well appear on national television as some kind of dainty-fingered fraidy cat.

I cautiously pick up the bits of wood with my fingertips, or nudge them over with my toes. The reason for this is simple: while I have never heard of a fatality caused by cockroaches, huhu grubs or slaters, I am leery of the whitetail spider. Its bite is far worse than whatever noise a spider may make.

Don't get me wrong, I'm no arachnophobe. I used to keep spiders as pets. I liked having them around to catch insects, and I enjoyed watching their webs develop into increasingly elaborate structures over a period of time. I would quietly approach them to marvel at the size of the beast that dwelt within. But where I saw fascination with nature's varied marvels, other people saw bad housekeeping.

The whitetail spider is different. It's a brash interloper from Australia, a place that seems to exist only to evolve creatures whose specialities are lurking and inflicting pain. Dangerously fast and fearless, the whitetail seems to have no qualms about running at you rather than scurrying away.

Considering that the whitetail eats other spiders, it does little to endear itself to me. Not only is it a nasty creature in and of itself, it is also throwing off the natural bug control system that we take so much for granted.

Shame on them.

A whitetail-inflicted injury is the thing I fear most on the farm. A broken arm or severed limb seems tame compared to a gross flesh-rotting wound. While there is a large body of evidence that says the whitetail's bite may not actually cause the festering wounds of rotting flesh attributed to it, I'm not taking any chances.

I've seen the photos of the injuries purported to be caused by some kind of spider bite. The wound generally begins with a red welt, and then within a few days becomes a swollen purple carbuncle that rapidly turns black and begins oozing rotting flesh. I can only imagine the smell.

Maggots may then need to be brought in and applied to the wound to eat the dead flesh in the hope of stemming the spread of the decomposition. I'm not sure which is worse — rotting to death, or having maggots attempt to stop the rot. I'm inclined to think that I would cleanse the suppurating wound with fire. It would be agonising, but at least it wouldn't be maggots.

My fear is compounded by having been nearly killed by a whitetail spider. I discovered one in my car one day as I was barrelling along the motorway, and had I not been so adept at shooing it out the window, I could well have perished.

I suspect that had I been killed mid-route, either by the spider's bite or by careening into the median barrier due to girlish spider-induced flailing, it would have caused much wonderment afterward as to how the accident happened. I could only hope that some sharp-eyed forensics type would have noticed the strange welts and drawn the right conclusion.

I have no idea how the spider got there. I was a little concerned that perhaps someone who I may have inadvertently wronged had placed it in my car. Was I the victim of a botched assassination? Do I know something I shouldn't? But what?

It is the thought of whitetail spiders that is foremost in my mind as I fossick through the piles of timber in an attempt to find enough useful lengths to complete my First Big Project.

I guess I should be thankful that the whitetail is the only terror. In the woodpiles of other countries there would be all manner of nasties, like snakes, centipedes or wombats. In reality, my greatest danger is being savaged by a rusty nail and contracting tetanus.

While life under canvas agrees with me, I am determined to make life easier and safer for myself by knocking up a sturdy shack. I need something into which I can install the potbelly stove in order to cook during the day, and to generate some warmth, lessening the number of hours I spend wrapped up in my sleeping bag. A shack would reduce the risk of a raging tent-shaped inferno.

It's a damp October afternoon as the crew and I all trudge into the field and stare at the spot that I have declared will make an appropriate shack site.

I have been led to believe (by she who shall remain nameless but whose authority I do not question) that I need no form of council permission to build anything that has a floor area under 10 square metres. The shed is to be constructed solely from materials scavenged from around the farm, and I am basing its construction around two central resources.

The first is the roof, for which I will use the fibreglass canopy that currently covers the firewood pile. While it is doing an excellent job of keeping the firewood dry, I believe it will do an even better job of keeping me dry. This will be the centrepiece around which the entire building will be constructed.

I intend to construct the framing of the building from the iron strips that currently make up the kitset shelving in the shed. They will function like adult-sized pieces of Meccano, and will be the ideal components for the structure.

I was a Meccano child. Not for me the simplistic plastic brickwork of the Lego set. I had the gift of nuts and bolts and pulleys and levers and steel strips and girders and plates and spanners and screwdrivers.

It is my belief that if you love your children, be they boys or girls, you give them Meccano. Learning basic engineering principles is a skill that will last them their lifetimes.

For windows for my shack, I will use several doors that have large panes of glass in them, which are currently lying against the wall in the chicken house.

All I have to do is make sure the posts are in the right place, construct and secure the steel framing, somehow get the canopy up onto the roof, scrounge some tin for the walls, fit the doors as windows, put in some kind of floor, install the potbelly, cut a hole in the roof for the chimney, then light the fire and relax next to it with a refreshing cup of tea.

How difficult can that be?

As the spade cuts the first sod from the earth, it occurs to me that I shall have my shack knocked up in a jiffy. As it turns out, I don't knock it up in anything approaching jiffy-ness.

This is my first introduction to the fine art of Building for Television. It's not for the hasty. Shots need to be lined up. Actions need to be repeated from different angles, so delays occur while the camera is repositioned.

As a rule of (an only slightly battered) thumb, I calculate that we will need to add at least a third more time to any job when a TV crew is filming it.

What is even more frustrating is that the whole elongated process of building will, no doubt, be compressed into a very short montage sequence. How they get things crammed into a weekend on those reality building makeover shows I'll never know, but what I do know is that I will never submit any property I own to one of them.

As we progress, there are long, animated conversations on the necessity of various pieces of safety kit. Should I be wearing goggles when hammering? Surely not, I think — the idea of a nail hitting my eye seems ludicrous.

I continue to think this no matter how many times a nail strikes me in the face over the course of the next few months.

Do I need gloves? A hard hat? A coloured safety vest? These

are not questions that face the amateur builder to quite the same degree, but seeing as we are doing these things on television, the crew consider that I should be seen as setting an example to the nation. I will be a role model.

I haven't considered these ramifications. The crew argue that viewers will assume that what I am doing is the correct way to do things.

I don't tend to do things the correct way. I do them my way. But it will make things difficult if we include the 'Don't try this at home' disclaimer for a series that is supposed to encourage people to emulate me in becoming more sustainable.

So I put on the goggles and the earmuffs, and wear the coloured safety vest for a while. It is quite a fetching ensemble.

I set to and begin to dig the holes for the poles. The poles aren't poles as such; they are simply the best-looking fence posts I can find.

After digging down a few inches, I find the topsoil gives way to clay. Orange and sticky and heavy, it clings to the spade, to my boots, and to the posts that I use to try to knock it off my spade and boots. It is so waterlogged that digging is virtually impossible, and the holes fill up with water faster than I can drain them.

As a result, the posts aren't planted quite as deep as they could be. They end up at the minimum depth I think I can get away with in order to make it look as if they are deep enough, without resorting to cutting anything off the top of them. I am fairly sure that the weight of the building will hold it down so that it won't blow away.

Then comes the simple task of bolting together the lengths of shelving supports to create the framing.

Everything is progressing well, until it is time for the roof raising.

Sometimes my four-dimensional visualisation skills let me down. I had planned to lift the roof on, and then try to turn it over to fit it. Thankfully, Shawn is on hand to see me right, and he politely points out that what we actually need to do is turn the roof over so that it is the right way up before we put it on.

Sometimes, when someone points out the correct way of doing something, it is almost impossible to conceive how you managed to even contemplate doing it any other way.

With a stiff breeze threatening to blow the roof away, Shawn and I, ably assisted by his stepson John, raise the roof.

All that is needed is for me to bolt the roof on before the increasingly gusty wind blows it off again. Fortunately, I have an abundance of bolts. Unfortunately, I haven't thought to ascertain the length I require. None are long enough. Leaving Shawn and John clinging to the roof to hold it down, I scuttle off to see if I can find some longer bolts.

By the time I return from failing to find any longer bolts, Shawn and John have the roof tied down with strops, and look as if they are sharing some kind of private joke with the crew.

'Job well done,' I say.

'Not quite,' responds Jane, looking at the vacant areas where walls and floor should be.

'Oh yeah.'

I crack on.

One of the great things about working for Jam TV is that they choose their crews well. Everyone pitches in. When we're not filming something, the cameras and sound gear go down, and hammers and saws and shovels are picked up.

Within short order, the doors are fitted as windows, tin is nailed to three of the four walls, while the fourth is constructed by the judicious use of a garage door left in the barn. Then the pallets are laid on the floor, Jane lays a little carpet on them, a large concrete paving stone — upon which I'll put the potbelly — is positioned in the corner, the potbelly is placed on that, and above it I cut a hole in the roof through which to put the flue.

After two days we are now nearly done.

As I couldn't find enough flue lying around, Bevan the cameraman, Frank the soundman, and I use all of our combined ingenuity to manufacture one. We take a length of tin, roll it around a post, use trailer tie-downs to ratchet it into a circle, and pop-rivet it together to create a new chimney section.

We stare at it in a chuffed manly fashion, before filming the moment when I drop it into place. Then it's simply a matter of filling the stove with wood, lighting it, and relaxing in its warming heat.

Not a bad few days' effort.

I have built my shack. I am sure my forebears would be proud. There are only two minor problems.

The first becomes apparent the next time it rains. It appears that I am suffering a rather nasty case of Leaky Building Syndrome. Water pours in around the chimney, running down the metal to sizzle on the hot potbelly. Quite a lot of water. I can't figure out what to use to block up the gap that won't catch fire due to the heat from the chimney.

The second problem is that it appears that I have broken the law.

While building something under 10 square metres in area is exempt from a building permit, the building still has to be built according to the strictures of the Building Code.

On a wet October morning, a friendly man whose job it is to inspect these things on an independent basis points out the obvious defects.

He suspects the poles aren't deep enough.

He isn't sure the shelf framing makes for the most suitable building framework, but I think he is impressed at the ingenuity of its usage.

He points out there is what he politely describes as 'a little moisture' leaking in from the windows that appears to be rotting some of the timber.

I point out that it isn't my building skills that are at fault here, but that the wood was rotten before I used it. I also point out that, to be fair, I have strengthened it with some of the spare shelving brackets.

Worst of all, he says, 'There is no way a potbelly can be installed without the sign-off of an engineer.'

'What kind of land is it,' I demand to know, 'where people can't inhabit a shack of their own construction? What would our ancestors have thought, in their hovels and shanties? It is an outrage.'

If my home-made shack falls down, I continue to rant, surely it'll be on my own head? Even if that's literally?

He agrees, and suggests that if I remove the illegal potbelly, promise not to sleep in the shed, and to build everything else to code, then he'll allow the structure to remain. As the cameras roll, I assure him I shall.

'Would you,' I ask, 'perhaps like to help me remove the stove?' That way he can be sure it's gone. 'I'd love to,' he replied, 'but, alas, I have other pressing council callings.' It seems that although building inspectors are to be feared, they are not as frightening as whitetails.

6

The elusive eel

Given my propensity for appearing in public in all manner of scanty attire, it may be difficult to tell that as a younger man I suffered from a self-diagnosed form of the psychological condition known as body dysmorphia.

In a world obsessed with the tanned body beautiful, I loathed taking my shirt off or wearing shorts in public, because I was a pale ginger-haired boy whose skin refused to tan, meaning I was either ghostly white or, after a little sunshine, hot pink — literally. As a result, I detested summer and everything that went with it. What I hated the most was that I missed out on a lot of swimming.

It could be said that I am now overcompensating for this childhood trauma by taking most of my clothes off as often as possible. I once suggested to someone that this might be a form of aversion therapy. They laughed and said, 'Yes, particularly for any women in the vicinity.'

Little wonder then that I decide it is entirely appropriate to illustrate the lifecycle of an eel for an insert in the show by donning a giant eel head and, alternately, my scanty swimming trunks and a bikini. I like to think that the visual metaphor is fitting, because for many people, both the eel and my good self in a bikini are visions that are slightly uneasy on the eye.

Eels are one of those animals whose lifecycle is difficult to understand, not only because little about them is understood, but also because what is understood seems slightly ludicrous.

Over the millennia, some of the greatest minds in history have tried to decipher the mysteries of the eel.

The Greek philosopher Aristotle, when not mooching around being profound and inventing thinking, studied the eel. He declared that they were born of the earthworm, and emerged literally 'from the guts of wet soil'. Given the rest of his accomplishments, I guess we can forgive him this one minor error.

Later, Sigmund Freud began his career dissecting eels for several weeks in an attempt to find their testicles. He failed to do so, and some wonder if his later sexual theories were somehow caused by his frustration in this endeavour.

It's astonishing to think that the eels that inhabit the creeks and ponds of deepest, darkest New Zealand were all born thousands of miles away in warm tropical waters. No one knows exactly where, but it is being narrowed down to an area somewhere near New Caledonia, or Tonga, or maybe Fiji.

Quite why they would choose to leave these lovely tropical waters for the dark pools of innermost New Zealand I do not know. Nevertheless, here they remain for up to 40 years, and in some cases 80 or even 100 years, before making the return journey to the spawning grounds. It seems an inordinate amount of effort to go to.

No one has ever observed eels spawning, but I can only imagine that it is an absolutely astounding sexual experience, because as soon as it is completed they die.

As fascinating as all of this undoubtedly is, the thought of having to deal with one is not in any way appealing to me. I've never liked them. They give me the willies. I think it is a combination of their slimy writhyness and their teeth.

That they can grow up to 1.75 metres long and weigh up to 24 kilograms means that I have no trouble believing the anecdotal reports of ducklings and other small waterfowl disappearing underwater as they paddle along.

I have no desire to be bitten by an eel, as their teeth are set at a backwards angle so that when they bite down, their teeth lock in place, and the only way to release them is to basically tear out the flesh they have gripped. That has to sting.

This is why I am prepared to risk the mocking laughter and snide glances of Shawn's stepson John and his mate Brin, who have somehow been persuaded to take me eeling, when I pull from my pocket a sturdy leatherette man-glove that I have brought with me in case I am forced to handle one of these repulsive, but apparently tasty, denizens of the creek.

What makes the entire eeling endeavour all the worse is that the eels are to be tempted by a better quality of meat than I have seen in several days. The bait consists of a large hunk of wild venison that John has taken from Shawn's freezer. What I really want to do is take the bait home and simply eat that instead.

As Jane won't let me, I find myself casting a line into the creek, forlornly hoping that I will not catch an eel. John and Brin though, not wishing to disappoint, have what they proudly refer to as their never-fail eel-enticing trick. This consists of cracking an egg into the creek.

As they proceed to crack egg after egg and drop them in the creek, they insist their technique remains a secret. I assure them it will. To date, I have told no one.

Whether it is the lure of the eggs or the venison, within a short period we are hauling eels out of the creek. I insist on keeping only one — purely in the spirit of sustainability, of course.

It is rather difficult, in my experience, to tell if an eel is actually dead. As they can absorb up to 50 per cent of their oxygen requirements through their skin, merely leaving them out of the water is not enough. You can stomp on their heads, but this seems barbaric, and there is always the danger that the soft ground will mitigate the stomping and you will merely end up with an angry eel with a headache.

Aiming a knife at their brain is another method, but even then it can be difficult to tell if the brain is stabbed, as they will continue to twitch and writhe in an inelegantly disturbing manner for some time.

I suggest that John employs both of these methods in an effort to render my eel as dead as possible. I even offer him my man-glove, which he politely declines.

With the deed done, I pack the eel in a sack, bid the youths farewell, and set off to feast in the manner of a pre-European Maori.

The first step in preparing the eel is to de-slime it. John has suggested a couple of options, including putting it in a bag of salt overnight. As they are to be my dinner immediately, I need them unslimy pronto. I decide to employ the dipping-them-in-hot-water method.

With my delicate hand safely enclosed in the leatherette gauntlet, I tentatively grasp the heavy body of the dead eel and drop it into a wok full of steaming water. The dead eel immediately begins such a frenzy of quite undead writhing that it pops right back out of the wok.

While I am sure it is dead, the water must have caused its nerves and muscles to expand and contract spontaneously. This is not family viewing.

'I suspect we won't see that bit on the telly either, will we, Jane?' I say.

'No', she agrees. 'We most certainly will not.'

The eel finally stops wriggling after I hold it down in the boiling water with tongs, which doesn't please me entirely as it means that I now have to dissect it.

This is not overly difficult, as one simply needs to slice it from what appears to be the opening of the anus, up the torso, to the neck. The intestines and other organs can then be easily removed.

As it is nearly dark, and the crew are due to knock off, I decide that, rather than fillet the critter, I will cut it widthways into steaks, not dissimilar to salmon steaks, and fry each with a different coating of herbs I have scavenged from the garden.

So, with the eel steaks slathered in a little olive oil, I coat one in rosemary, another in chives, and the last one in garlic, before hiffing them into the pan sitting atop the gas cooker.

I suspect there is a reason that I have never seen eel offered like this in a restaurant. That reason may be that the result is not dissimilar to slightly fishy-tasting rubber.

But, on a cold, damp October night, as the crew disappear into the darkness, that fishy-tasting, herb-flavoured rubber proves to be warmingly nutritious, and a thoroughly welcome break from silverbeet and eggs. As I pull bits of its stringiness from out of my teeth, I try not to think of the venison.

7

A can of worms

It is decided, by those who make the decisions, that I am to become a worm farmer. This is an exciting development because, prior to this, I have only ever grown worms internally.

The worms will become my untiring minions, working relentlessly to convert food scraps and other organic matter into vermicast or, as it is also known, worm castings. It is, in its most crudest sense, worm faeces.

As the worms slurp their food into their little worm-mouths, it passes through them and into their stomachs, which teem with all manner of bacteria and enzymes. Not only do these microscopic creatures help with digestion, but many of them emerge in the castings, and when added to soil enable the release of nutrients that would otherwise remain locked in the soil.

Or at least that is what I read on a packet of the stuff available in a garden centre for only slightly more money than I have on me.

The process also produces a liquid, known commonly as worm tea, which when diluted with water gives plants a good kick in the roots.

And the best thing about worms is that they can be ordered over the Internet. Do livestock come any simpler?

I had first seen worms employed in a large-scale way on my tour of Rainbow Valley Farm with Trish.

She had lifted up the lid of what appeared to be a coffin to reveal a layer of food scraps, which she scraped away to reveal thousands of worms, all wriggling around, doing whatever it is that worms do.

I was just a tad reluctant to accept her offer to plunge my hand into the rich humus they had created. It is, after all, worm faeces.

Trish seemed to know an awful lot about worms.

'They eat their own bodyweight in food every day, are both male and female, and have up to two hundred babies a year. They have five hearts and no teeth, so they grind up their food in their gizzard, which contains tiny stones,' she told me while holding up a handful of the striped squirmers and looking at them with what appeared to be a sense of reverence.

As someone who was contemplating farming worms, I really only had one question: 'If I have a worm farm, and I'm a little short on worms, can I chop them in half and double their numbers?'

'No, that's a myth,' she said.

'Oh. That's a shame.'

For my worm farm I have acquired a disused cast-iron bath, in the bottom of which is painted a rudimentary mermaid, complete with long hair, a fishlike tail and ample breasts. The simple drawing makes it appear as if someone has killed the mermaid in the bath (probably not by drowning), and then the police have drawn an overly elaborate and mildly erotic chalk sketch of the outline of her corpse.

In order to harvest the worm tea from the bath, it will have to be built high enough off the ground so that I can fit a bucket underneath the plug hole to collect the juice.

My plan is simple. I'll firstly dig four posts into the ground at a depth that will be determined by making an educated guess as to the minimum depth the holes need to be in order to keep the posts upright, divided by how energetic I am feeling at the time.

I then plan to nail cross-beams onto the posts to create a box-shaped frame into which I will drop the bath so that it rests on the beams. The worms will live in the elevated bath, the tea will drain into the bucket, and all I will have to do is remember to feed the worms and empty the bucket onto the garden. The plan, as I formulate it in my head, seems like a simple task that should be easily accomplished. What I fail to factor in is the reality of the bath.

There isn't much that is as clumsily heavy as a cast-iron bath. While I can roll it over with a modicum of difficulty, I haven't taken into account quite how I will get it up onto the platform without being crushed or rupturing myself.

I have assumed that the endeavour will be filmed in sections, as it usually is, and that in between shots the crew will assist me in the task, as they usually do.

Jane has other ideas. She wants the entire effort filmed with a locked-off camera shot. This means that I will have to complete the task alone as the crew loll about behind the camera, watching me struggle. The simple reason why she wants it done this way is that she is a sadist.

When one engages in activities such as farming, there are some jobs that are on the very margins of being possible to achieve alone. While it is a wonderful personal challenge to attempt a job that might be a little too much for one person, there is always the danger that during these little tests of ingenuity and strength, something could go really quite badly wrong. People can be hurt. People like me.

One consolation is that I am in the fortunate position of having a camera crew documenting my attempts at these outlandishly ambitious tasks. If I am hurt or injured, investigators will have the videotape handy to ascertain exactly where things begin to go wrong. I would suggest it was about the time I had the third glass of Astrolabe chardonnay at SPQR in August.

What is worrying about the prospect of failure is that the level of failure will be so ludicrous that I will become the laughing stock of the accident investigation and insurance industries. I feel that what I am about to attempt will one day be used as the flashback section from an episode of *Seconds from Disaster.* Such is the nature of our appetite for this type of televisual carnage that the same footage will no doubt also appear in *New Zealand's Funniest Home Videos.*

While the crew relax in the sun, the unblinking eye of the camera stares fixedly at me while I stare at the bath.

Performing a task like this gives you a good sense of what sportspeople must go through in any of the trials involving the throwing and lifting of heavy objects. They are the most individual of events. It's you versus gravity. Mind over matter. Of course, matter does have the advantage of physics.

Stepping forward, I crouch, wriggle my fingers under the lip of the bath, and clenching my sphincter in an attempt to prevent any kind of exertion-based prolapse, muster all the strength I have in my feeble legs and propel myself and the tub in an upwards direction.

As I realise I am lifting the bath successfully, a primal 'Harrumph' of manly exertion and triumph emanates from somewhere deep inside my face. With another Herculean heave I slide the bath along the railings so that it ends up sitting in the right place, albeit upside-down.

That wasn't so difficult. All I need to do now is flip it over and lower its heavy mass gently into place.

I manage to roll the bath onto its edge, and as it teeters precariously, I realise that my chances of slowly working my way around it, while holding it up, then allowing it to lower gently into place, are at their best slim, and probably closer to nil. Oh well, I think, I'll just let it fall into place.

I let go of the bath, leap deftly back as gravity takes charge of it, and watch as the bath clangs heavily but perfectly into place. I keep watching, as the weight of the falling bath then knocks off the boards intended to hold it up, and the boards, the bath, and my ego all crash heavily to the earth.

I stare at the fallen tub with the look of a man who hasn't really anticipated this particular outcome firmly plastered on my face.

The crew look at me with the expressions of people who are not overly surprised at this particular outcome on their faces.

I look up at the crew.

They continue looking at me.

Then we all laugh.

It couldn't have been scripted any better if Benny Hill had written it, although if he had, there may have been more women in bikinis present.

What I need is more nails. Actually, what is needed are bolts, but I don't have any of those, so more nails will have to do the trick until I can wangle some bolts.

As I proceed to alternate between hammering more nails into the structure and whacking my thumb, Frank the soundman says, 'You know, people used to mock Maori for trading land for nails.'

He is right, of course. The belief that Maori were duped into trading any manner of things for a nail instead of money is an easy concept to mock, until you try building something without a nail. You can do without a hammer, but it's clear why a nail would be considered a taonga.

At the time when the first contact was made between the Maori and Europeans, nails were still hand-made, and things of value, even to Europeans. It was reported in the 1770s that large nails, which could be turned into chisels, were the most valued trade item when dealing with Maori. Small nails were fashioned into fish hooks.

It's not until you don't have access to nails that you stop taking them for granted. Nowadays we cast a bent nail aside. They are just another seemingly disposable object. The process of pulling them out of timber and straightening them is considered too time-consuming.

At one stage I wonder, given the meagre nature of my stipend and the challenge of becoming more sustainable, about the possibility of melting down used nails and re-forging them. Sadly, this idea will go the same way as the ideas I had about a bicycle-powered shower, the aqueduct, and the pig-poo methane extraction plant — all of them falling somewhere short of the drawing board.

With the timber now held up by stronger nails, I heave the bath back into place, flip it over, and again in fear of being crushed, release it to the bonds of gravity.

It slumps into place. I hold my breath. Nothing moves. I stand still for a few more seconds, waiting to see if anything is about to become detached.

Nothing happens.

Blessed nothing.

As I place some netting into the bath that will allow a space to be retained between the bottom of the bath and the vermicast so that the precious worm tea can dribble out, and cover the netting with cardboard, scraps and worms, the mermaid becomes obscured.

Realising that she will soon be lost under vermicast, I become a little sad. No doubt she brought someone great joy at some time. It would be nice to know who drew her, and why. Did she have a name?

As a final act, and because worms are like goths and prefer the darkness, I place an old door over the tub like the lid of some kind of rudimentary coffin, and think to myself, 'Job well done.'

Now, all I have to do is remember to feed them.

8

Heard of goats?

While I said at the outset of this adventure that I don't mind getting up early, what I really meant is that dawn is a special thing, and with it being so special you don't want to ruin its magic by seeing it too often.

There is something both a little exciting and oddly wearying about rising at the ludicrously early hour of 3.00 a.m. It's a long time till dawn. Nonetheless, I rise at that time so that by 4.00 a.m. on a chilly, late spring morning, I can head up a tanker track towards the lights of a cowshed in a location I cannot reveal, as Chris, the owner of the property, fears rustlers.

He's worried that they will come after his prized milking goats, which I shall be assisting him in milking in exchange for anything that isn't eggs or silverbeet or grapefruit.

I am feeling quite chipper after the hair-raisingly brisk ride on the quad bike, which is a marvellous workhorse when it's not rolling over on top of its rider. The eyes of hundreds of goats sparkle in its headlights, like blinking, yellow stars, as they cluster around in the chill dark of the morning, ready for milking. With the gate opened they flow out like a river of tiny, terribly placid goats.

They plod past us as Chris explains that he never intended to become a goat farmer, but he couldn't afford a herd of cows. Seeing the goats advertised, he thought: why not? I will think the same thing as the milking progresses, and the goats begin to enchant me.

I have been defecated on and urinated on by more cows than I care to remember as I stood beneath them in my parents' cowshed. Often it was a few moments of welcome warmth on a cold morning, tempered by the freezing water that then had to be used to wash the slurry off. After a while it merely became an inconvenient annoyance for the hapless person underneath, and a source of earthy humour for everyone who witnessed it, which in my case was usually my mother.

But with the goats, there is no such problem. They are a clean animal. Rather than the voluminous torrent of reconstituted grass that can pour forth from the cow, the goat emits only tiny little pellets.

If I were to return to a dairy farm, I would have no hesitation doing so with goats. They are small, easy to handle, and delightfully cute.

Placing the cups on their two teats is a simple task made difficult by my almost ingrained four-teated cow habits. Goats are also wonderful producers, and it's hard to imagine why more people don't wish to milk them. It's like novelty milking.

'The white ones are for whole milk, the brown ones provide the cream, and the black ones produce skim milk,' Chris tells me.

'Really?'

'No,' he says. 'I'm joking, now hurry up and drench them or we'll be here all morning.'

Even though they are all nanny goats, they have little beards under their chins, which makes them easy to drench, as you can gently grip it and turn their heads to the drench nozzle.

As I'm hosing the yard down afterwards, the rising sun's rays catch in the water, creating a silver stream that rhythmically washes away the goats' leavings. It's an iconic New Zealand scene, one that can often be witnessed from cars as you travel through the countryside in the early morning.

That watery parabola is hosing away what is also one of dairying's biggest problems: cow effluent. Where once it simply went into slurry ponds to settle, with much of the overflow often finding its way into drains and waterways, it is now much more regulated. And a good thing this is too. There are even those who argue there's money in muck.

I met a scientist in Palmerston North by the name of Dr Steven Pratt, who is endeavouring to turn cow excrement into bioplastic. He showed me a bucket containing his research, which contained a slurry, on the surface of which it was possible to discern a faint, oily gleam that I was told was the beginning of the bioplastic.

He talked passionately about all manner of scientific things, such as 'the inclusion of propionic acid producing polymer chains with the favourable characteristics of malleability,' which, despite sounding fascinating, made no sense whatsoever to me.

It did seem clear to me, though, that there is a high chance at some stage in the future that cow effluent could become a renewable, biodegradable source of plastic. It doesn't get much more renewable than cow manure. If it proves successful, it may be that the grass is actually greener after it has passed through the cow.

While Chris heads off for breakfast, it's my duty as the lackey to take the last of the goats back to the paddock. Some of them are a little lame, so they get to ride in the motorbike trailer as I chauffeur them back.

There's little I find more pleasurable than puttering along behind stock in any of the glorious rural landscapes we are blessed with in this country, with the sun just risen and the whole day stretching before me.

Everything has gone well, and all I have to do is to take the last goat from the trailer. As I lift it out, I assume it will, in a nimble example of goat agility, simply bound out. It doesn't. It falls out.

It could be argued that I dropped it on its head, much to the amusement of Bevan the cameraman, who barely manages to capture the shot as he is laughing so much. The shot will never make it into the show — yet another example of something that is not family viewing — which is fortunate, as it may have destroyed any chances I have of working on a goat farm in the future.

With the goats safely back in their paddock, it's time to head home to my own chores. The wonderful thing about getting up so early is that it leaves a lot of the rest of the day to fill. I am to fill mine with more milking.

The ease of milking the goats is tempered by the milking of my cow. As neither Harriet nor Coco is in milk, I have bought another, named Judy, along with her two precocious stepsons, who she is milk-mothering. They are Hereford calves, one ginger and the other black, and somewhere along the way Jane has named them Sainsbury and Campbell.

The irony is that I got into comedy and television to avoid a career in farming. Well, actually I got into it because my promising law career was stymied by an inability to pass the required exams, leaving no other option but television, the last bastion of those who aren't really sure what else to do.

I grew up on a dairy farm near Huntly. I left the farm not because I hated farming, but because I hated milking cows. Specifically, I hated milking them in the heat of a summer's afternoon. I would begin to get a palpable sense of horror as the appointed hour approached.

Now, here I am, milking my own cow.

Admittedly it is a far cry from milking as I knew it. Rather than using machines as we did on my parents' farm, I have to milk Judy by hand, which is a time-consuming, laborious job that requires a lot of patience and skill.

I really should have taken the calves off her the night before; I had no idea they drink so much.

I don't normally need much milk, so I only milk her occasionally, but on this particular day I am supposed to have several litres in order to make cheese with renowned cheese-maker Sabine Drueckler.

I don't have several litres. I have half a cup.

Sabine lives at the Otamatea Eco Village, an 'intentional permaculture community', situated on 250 acres on a peninsula of the Kaipara Harbour.

'I've had a bit of a milk disaster,' I explain as I arrive at the village, offering my cup of milk up for inspection.

Sabine peers into it and says in her Germanically accented English, 'This would make cheese about the size of the tip of my thumb.' Luckily her husband, Wolfgang, has persuaded their cow to give him five litres that morning.

She suggests that I pour my offering into the bucket that her strapping husband has diligently filled. This way, it will not be discernible which bit of the milk is mine, so I will be able to lay claim to the entire cheese. Who could argue?

'Today we are making feta, so we begin, yes?'

'Yes.'

I had not realised that making cheese is such a simple process. We filter the milk through a small sieve, heat it up to 40 degrees — the same temperature as the inside of the cow, Sabine tells me — then add some home-made yoghurt and a little rennet to set the ball rolling. Then we have to wait for an hour.

This is the part of the procedure at which I excel.

After an hour Sabine announces, 'It is time now to cut the cheese.' Because I am immature I find that very funny, especially as it is delivered with a slight German accent.

By this time the milk has solidified, and appears to resemble an instant pudding. With a long knife we cut it into tiny squares, which allows the solid curds and liquid whey to separate. Draining the whey, we take the curds and press them under some cans.

As I can't wait the 24 hours needed to complete the process, Sabine performs the trick perfected by television cooking shows the world over and produces a cheese she prepared earlier. It resembles a block of polystyrene.

'Tasty polystyrene though,' I hurry to reassure her.

It is that simple. We have a friend in cheeses.

On the way home, I think that if it is that simple, then perhaps I should do more of it. After all, even making butter is easy; it just requires a very strong wrist and a lot of shaking. The only drawback is the milking process. If I am going to do it in future, I will need to extract more than the pitiful cupful of milk I got from Judy's udder. There has to be a better way, and, as it turns out, there is.

While at Dexcel's Greenfield Project, a research farm in the Waikato, I witnessed the brave new world of dairy farming. I beheld the marvel that was the robot milker.

It was a little disappointing.

While I certainly wasn't expecting an android bedecked in the clobber of a busty milkmaid, yodelling as it worked, nor was I expecting a big grey stainless steel box. The cow walked into the box, a scanner identified her, a mechanical arm containing the cups swept out underneath her, sensors based on similar technology to that used in carwashes located her teats, independent arms applied each of the cups, and as she milked out, all manner of data about production was logged.

While this happened, the cow stood calmly eating a special brew of food and supplements doled out to her particular specifications by the feed machine.

As the scientist remarked, 'It doesn't matter how technical we get, only the cow can produce the milk.'

The cows were also being used to trial the reduction of farm labour even further, by being allowed to dictate the times they wished to be milked. This meant they could walk from the paddock, be milked, and then be directed back to the next paddock via sets of automated gates.

This was dairy farming at its best, a glorious devolution of authority, with it now being vested in the best workers the farmer has — the cows themselves. No more would the hapless cocky have to rise in the predawn dark, or rush back to the cowshed from a family do in the afternoon. The bovine sisters were doing it for themselves.

If I had been able to utilise the services of a robot to milk my parents' cows, I may well have entertained the idea of remaining a dairy farmer.

I can't even remember if the robot had a name. No doubt it would be something like Trev.

'Is there nothing you can't teach a cow?' I asked the scientist as I was leaving.

'Well, we don't like to tell the farmers because they might complain,' she said, 'but we're trying to figure out if they can read.'

'Books?'

'No, signs with coloured symbols,' she said.

That's good, I thought. You wouldn't want a cow reading the newspaper — she might find out what she's worth to the farmer and demand more molasses and a back rub.

9

Sheep in sheep's clothing

Two important things happened in New Zealand in 1971. Firstly, I was born, and secondly, the Agrodome opened in Rotorua.

Over the years, countless millions of tourists have been mustered through its hallowed hall to sit open-mouthed in astonishment at the exploits of one this country's greatest ever animals — the sheep.

I often find myself staring at my sheep in open-mouthed astonishment. Any time I venture anywhere near them, their eyes widen, their breathing increases, and they begin to look decidedly agitated. After a few moments, they flee in the opposite direction as fast as their tasty lamb shanks will carry them.

I think they may have had psychological issues. They remind me of a group of teenagers who have been up all night watching horror movies and then suddenly think they may have just heard something tapping on the windows.

Because of their emotional instability, and the fact that they are destined to end up in my stomach, I generally leave them to their own devices in the paddock I call 'The Tops'. It's a nostalgic and slightly self-aggrandising name that pays tribute to the great high country sheep stations that run up into the Southern Alps. In reality it's just the paddock at the top of the hill.

Prior to emigrating to New Zealand, my great-grandfather was a shepherd for Queen Victoria at one of her London farms. I doubt he had a close relationship with her. It's unlikely that as Her Majesty passed in her carriage she would lean out of the window, wave, and holler, 'Hello, James, how are the sheep?' and he'd doff his cap and shout, 'Mighty fine, Ma'am,' and she'd reply, 'Oh that's grand, why don't you pop over later for a cuppa?'

None of his skills with sheep seem to have been passed down to me.

The only time the sheep's erratic behaviour is of any concern to me is when I need them to be somewhere specific, such as in the back of the Land Rover. The reason they need to be in the back of the Land Rover is that they need shearing, and the shearing shed is some distance up the road at Shawn's.

While the thought of mustering a flock of sheep along a country road conjures up classic images of a rural New Zealand reminiscent of iconic Marti Friedlander photographs, the reality of running my four unpredictable scatterbrains along a busy modern highway is quite a different story.

Hence the need to catch them.

So it is that Shawn, his four-year-old daughter Charlotte, his dog Bob, and I find ourselves all slowly pirouetting in the middle of the paddock as we watch the sheep gallop around and around and around the fence line.

With the help of the dog, we finally corner them in the shed, and now all we have to do is grab them. I can well understand why rural areas produce rugby players who are such proficient tacklers. When it comes to dealing with an animal as mentally unpredictable as a sheep, you can't pussyfoot around. You have to commit yourself totally to the physical action of catching them. If you don't, and they get away, immense frustration ensues.

Inching closer to our woolly foes, Shawn leaps at one and in a nimble movement catches it and simultaneously flips it over, enabling him to drag it to the truck. This causes one of the others to make a dash for it, and as it runs past me I leap at it and am dragged halfway along the shed before my weight causes it to come to a reluctant halt.

'Got it?' asks Shawn.

'Yep,' I say, as I try to casually flip it over, an action that results in a flurry of flailing legs and near-catastrophe, until finally it is on its back and I have it firmly caught, much to the disappointment of Shawn, his daughter, and the crew, who are all enjoying the show.

I manoeuvre it to the Land Rover, hoist it up inside to join the one Shawn has just put in, and close the door.

'Well, halfway there,' I say.

Once again Shawn snares his with ease, but the sheep I'm after easily eludes me, and is headed towards the only gap from which it can escape at breakneck speed. All that stands between it and freedom is Charlotte.

'Grab it,' I yell, before realising that she is a four-year-old child who is probably not equipped to deal with the large and terrified mass of mutton that is about to run roughshod over the top of her.

All I can think is that I'm going to be responsible for breaking Shawn's child.

With a leap that can barely be described as valiant, I launch myself at the sheep and pull it to a halt just in front of Charlotte.

'I wouldn't have let it escape, silly man,' she says.

I believe her. She's her father's daughter.

The sheep are now secured in the back of the truck where they stand looking confused, their eyes hanging out nearly as far as their tongues. I clamber in, hoping they won't decide to leap over the back seat into the front to join me on the short drive to Shawn's.

I've only ever shorn one sheep prior to this, and that had been under the tutelage of the man who almost single-handedly revolutionised the shearing industry.

Ivan Bowen was a sprightly 91 when I had the great pleasure of interviewing him for the documentary series *Homegrown* at the

Agrodome, which was founded by his brother Godfrey with George Harford, a local farmer.

Together, Godfrey and Ivan developed a shearing style that was based on keeping the sheep as relaxed as possible, and which streamlined the movements required in removing the fleece, eliminating any unnecessary actions. It became known as the Bowen technique.

Despite the fact that humans have been harvesting wool from sheep for millennia, this breakthrough was revolutionary. When Ivan and Godfrey went to the United Kingdom, they raced the local shearers, who claimed the New Zealanders must have been cheating, as they believed it was impossible to shear so fast.

Such was the impact of it that Godfrey, a born showman, became an international celebrity, jet-setting around the world to demonstrate his shearing prowess in front of royalty and heads of state. He even appeared on Johnny Carson's *The Tonight Show*, where he sheared a sheep in front of tens of millions of no doubt astounded Americans.

The Soviet Union was so impressed with him that in 1963 he was honoured with their highest awards, the Hero of Labour and the Star of Lenin.

Competitions sprang up around the country as shearers vied to better one another; they became recognised as sportsmen rather than simple labourers. The most famous competition of them all was the Golden Shears. In its inaugural year in 1961, the crowds were so overwhelming that the army had to be summoned to maintain some semblance of order.

The Golden Shears and the shearers who competed in it soon became household names, and tickets had to be booked a year in advance. The shearers themselves became heroes, chasing ever-larger prize money, and were much sought after by sponsors and companies wishing to use them to promote their goods. Some shearers even began adopting training regimes, much to the horror of the traditionalists.

It may be hard to believe now, but sheep shearing was at the time

New Zealand's equivalent of rock'n'roll. Quite what this frenzy says about the state of entertainment in New Zealand in the 1960s I do not know, but it's little wonder that, to this day, people think of us as a nation that is obsessed with sheep.

Shearing had come out of the woolshed and onto the stage, and there was never a better time to open the Agrodome.

Ivan became the senior showman there in 1972, and so committed was he to his task of extolling the virtues of the sheep that even after a fire destroyed the original Agrodome building in 1980, he performed the show on a nearby hill in front of 100 visitors, while the building smouldered in the background.

Though Ivan finally retired at the age of 73, like so many of his generation it seemed that 'retired' wasn't a word he really understood. New showmen were said to quiver in their little shearing moccasins whenever he cast a stern eye over the sheep they were shearing to make sure his rigorous standards were being maintained.

When I met him, he still had a twinkle in his eye, and the manner of a showman about him as he extolled the virtues of his beloved sheep. 'The sheep is the greatest animal in the world! Do you know why? Because they provide us with food and with clothes. No other animal is as useful as the sheep, busy twenty-four hours a day growing the king of fibres — wool!'

The patter flowed off his tongue with the smoothness of the fleeces he'd removed from countless numbers of sheep. It was hard not to be captivated by him.

This was a man who had shorn more sheep than many of us will ever see in our lives. At his peak he was a world-record holder and five times the New Zealand champion, and could shear a sheep in under a minute while blindfolded.

The year before I met him, Ivan had shorn a sheep on his 90th birthday in a mere 2 minutes 41 seconds. It took me that long to get one out of the pen so that he could instruct me in the subtleties of the Bowen technique.

He was a patient but demanding teacher, guiding me through the series of sweeping movements he and his brother had devised to separate the sheep from its fleece.

The sheep proved fairly cooperative once I had it flipped over, and seemed quite happy to lean up against the ball of my knee. Some psychological or physiological impulse renders them virtually stupefied in this position — even the wildest sheep will simply go limp and sit there, looking around. It's really quite strange.

As I followed Ivan's directions, my main concern wasn't that I would leave too much on the sheep, but rather that I would take too much off. I wanted the sheep to leave with all of the nipples, ears and pizzles it came in with.

The razor-sharp blades, or combs, the shearers use are positively lethal. That they are attached to a vibrating headpiece that threatens to leap from your hand at any second only makes it more complex. It's an ungainly instrument for a newcomer, and it's made worse by being joined by a small pivot at the end of the handle to a long straight tube down which the drive mechanism runs. I spent most of my time untangling myself from where I had somehow managed to wrap myself around it.

Several minutes after I began, a confused but mostly shorn sheep clattered down the exit chute and out the pen, and Ivan offered me a warm and sincere congratulations on my effort. It was one of the proudest moments of my life to accept the compliments of this living legend.

For the rest of the morning he regaled me with his tales before wandering off to spend the afternoon training a sheep dog, a hobby that was his passion until he passed away a few months later.

As I struggle to remove the wool from my sheep in Shawn's dilapidated shearing shed, it is clear that I haven't retained too much of the technique that Ivan had taught me.

Luckily, Shawn is a bit of a dab hand at shearing and he manages to guide me through the process.

It is an hour of unmitigated chaos.

By the time I finish, I have a pile containing most of the wool on one side of the shed, and four denuded sheep on the other. I say most of the wool because I have missed several large tufts that still cling to various areas of them.

'I reckon that'll be fine,' I say.

'If you think so,' says Shawn. 'They're your sheep.'

'Probably need to practise a bit more if I want to go to the Golden Shears, eh Charlotte?'

'A lot more,' she says.

10

A giggle of girls

A few years back I bought a live-in lady friend a bread-maker. Yes, I'm quite romantic. I knew at the time that I would soon be eating the most expensive bread I would ever have the pleasure of buttering. I based this on the eventual-cost-per-loaf ratio, factoring in both the cost of the machine and the ingredients, as I suspected that, given the number of loaves that would be produced before the novelty wore off and the hassle began, this cost would be high.

For a few glorious days the smell of hot bread filled the house. Then it was just on weekends. Then we would occasionally think about pulling it out when friends came over for irregular lunches, but there wasn't really time and we never had the right ingredients, and besides, we needed two loaves, and it was easier just to buy it.

Now it is that I find myself beside a fire with a contingent of Scouts, all kitted out in their Scout outfits, complete with scarves and woggles, making damper. It's about as far away from a state-of-the-art bread-making experience as you can get, because damper is the most basic bread there is. It's made from little more than flour and water, and occasionally baking soda and milk.

As they troop in, I notice one alarming difference between these Scouts and my time as a Scout. They are all girls. There's nary a boy to be seen, other than the bearded leader.

'Where are the boys?' I ask.

'They didn't want to come,' one of the young women Scouts says.

I hadn't expected that the Scouts would all be girls. In my day, Scouts were boys who built rafts, and girls joined the Guides and sold biscuits. They didn't don a scarf and a woggle and associate with the boys.

Times have surely changed.

I haven't had a lot of experience with young girls. When I was that age, I was firmly ensconced at boarding school, and never encountered them at all.

I have no idea what to say to them that won't make me look like a silly — or worse, creepy — old man.

The Scouting movement is considered to be a proto-environmentalist movement due to Scouting founder Lord Baden-Powell's urgings for people to get close to, and enjoy, nature, so it is fitting that they should join me in my sustainability experiment.

So sustainable are we that today our bread-maker is to be a stick.

I don't recall having made damper as a Scout. We went caving, built rafts, did confidence courses, and learnt how to tie reef knots and bowlines, sometimes while rafting through caves to increase our confidence.

Sitting with the Girl Scouts, with a wodge of flour and water jammed onto the ends of our sticks, and the ends of the sticks propped over the fire, then tearing steaming hunks of the stodgy bread off the sticks and eating it, I realise that I am probably having more bread-based fun with my stick than I ever did with the expensive bread-maker.

I wish I could share with them my favourite recipe for non-store-bought bread: beer bread. The yeast in the beer helps the bread to rise.

Beer bread satisfies three of the main things that are great about food: it is easy to make, it results in hot bread, and it has beer in it. Simply by mixing three cups of flour, three teaspoons of baking powder, and 300 ml of beer, then baking until done, you can have a thick, doughy loaf of bread, best eaten when fresh, but marvellous toasted.

I toy with the idea of selling a few loaves of it at the forthcoming Spring Fling, but by the time I make a decision I have drunk the last of the beer.

As spring arrives, the good folk at Frech Orchards Estate, an organic avocado farm at Shelly Beach near the South Kaipara Head, host their annual Spring Fling, where people gather to see their wares and enjoy bands and a day in the sun.

Jane has arranged that I will have a stall from which I will peddle my produce. The problem is, I don't really have any produce to peddle.

Surely, I think, there must be something that I can utilise.

Then someone mentions dandelions, which are considered by some to be a useful culinary and medicinal herb. If this is true, then it is possible that I am living upon a veritable golden-flowered gold mine.

Some frantic research indicates that the leaves can be used as a mustard-flavoured salad green; the roots, when cooked, can be ground up to make a sort of dandelion coffee; and the flowers can be eaten or used for wine.

Further research determines that the French call the dandelion the 'pissenlit' — which means 'to urinate in bed' — in reference to its diuretic properties.

How is it, I wonder, that I have never known this?

Boiling the leaves into a tea is said to be a good treatment for nervousness. I wish I had known this before the Girl Scouts had turned up.

The problem with selling dandelions as produce is that the plants wilt the moment I dig them up and pop them into pots. Hopefully they will perk up again.

The only other things I have in abundance are grapefruit. I could simply bag some up and sell them, but if there is one thing that reading the business pages of the newspaper has taught me, it's that those involved in primary produce need to add value to their products.

What better way to add value to the grapefruit than by juicing them? As I cannot say for sure how organic it is, I shall sell the resulting juice as 'free-range grapefruit juice'.

But I feel that I need something more. I need A Point of Difference. Something that sums up my new life on the farm — something organic and sustainable. What I come up with is the decision to introduce to the world my new green initiative — the bonsai lawn.

I have realised that many people don't have the same access to grass that the rural person does — people such as the apartment dweller, or people who live on the second floor. This cannot be good for their souls.

I decide that what they need is access to a lawn, and who better to bring the lawn to them than me?

I need sods of the best grass I can find.

It is more difficult than you can imagine to find a good piece of grass when you are looking for a small but perfectly formed portion. In each small square I examine there is a weed, or a blemish, or an area that it isn't growing properly.

I spend a long time on my hands and knees, looking for the ideal square foot of turf. Then, using the spade, I cut the sod from the ground, trim it to fit with a knife, slide it into an ice cream container, and water it.

Soon I have quite a few bonsai lawns. Each contains quite different types of grass. There is something for everyone — from clover to rye to paspalum. There are short-grassed bonsai lawns, longer-grassed variants, and even pastoral grass.

Amongst the trees at Frech Orchards, people mill around the market stalls, plying their wares, strolling in the sunshine, and listening to the ubiquitous South American pan pipers.

I set myself up under a little gazebo with a mounting sense of excitement. I have never sold anything before, and here I am with my own produce stall. Will today be the day that I revolutionise the concept of the lawn? Will the world be reminded of the wonderful properties of the dandelion? Will I make my fortune peddling free-range grapefruit juice?

The answers are no, no, and not really.

The bonsai lawns do become a talking point, and a useful lure to get people to buy the juice.

People are intrigued.

'Do they come with little lawn mowers?' one potential buyer asks.

'Not yet,' I say. 'I could arrange some scissors.'

The multitudes that flock past the stall agree that they are a good idea, especially as I pitch them as a gift either for someone you love, or alternatively for someone you don't really like.

Many agree that they are a great idea for someone in an apartment. But in the end it is a little like trying to sell . . . well . . . grass to farmers.

The dandelions are also a talking point, but sales of them are non-existent.

Both, however, help to shift nearly all the litres of juice I have juiced.

I've heard that some other chaps are doing well on the juice front too, only their juice has a little kick to it. The men from the Helensville Home-brew Club are snug in the garage, offering samples of their wares for a small donation.

This is convenient, as I have made a little profit from my juice, and purely in the interests of research, I think I should see what kind of brew it is possible to make at home. The answer is quite a lot of good-tasting booze.

Graham, a red-faced barrel of a man, and Rudolph, an effusive Austrian in what could be his sixties, are only too happy to share the collective produce of their fellow club members. They have a selection of wines, beer, liquors and spirits.

'Try the mead, it's the oldest wine in the world, made out of water, honey and yeast,' they insist, offering a small glass of the yellowy liquid.

In my mind, mead has always conjured up images of Vikings drinking it from cow horns, before stripping off their furs to run nude into battle. I am hoping this state of affairs won't arise this afternoon, but there are worse ways to spend a Sunday.

The home-brew club celebrates the indomitable human spirit of creating spirits. If there is anything that has not at one stage or another been turned into alcohol by humans, I would be very much surprised.

I have always considered home-brewing a little like home bread-making. It is a lot of hard work for something that can be bought in the store for a fairly reasonable price.

After sampling the club members' liquors and a few pleasantly rough spirits, they insist I sample the manuka beer. There is little as authentically Kiwi as manuka beer, which has been brewed in New Zealand since the arrival of Captain James Cook, who was no doubt a little shocked to discover that, prior to his arrival, the country was booze free.

Historians assert that prior to the Europeans landing on our shores, the only things getting drunk in New Zealand were wood pigeons gorging on fermented berries. Some claim that the Maori — especially the older women — would occasionally drink fermented tutu berry juice, but other than that there was no liquor to be had in our shaky isles.

Clearly something needed to be done about this, and in 1773 on his arrival in Dusky Sound after a journey of 11,000 miles and 117 days without sighting land, Cook stood on the poop deck of his converted collier on the morning of 27 March and addressed his crew.

'Men,' he said, 'good work with all the sailing and being cooped up. I believe that you all deserve a beer,' and before you could say 'this'll put hairs on your chest', he grabbed some yeast and assorted rimu and manuka twigs, biffed them in a pot with a little molasses, and brewed his shipmates a beer. And so began a great tradition of rank amateurs brewing beer in this fair land.

His crew praised the results, one writing that 'after a small amount of rum or arrack has been added, with some brown sugar, and stirred into this really pleasant, refreshing and healthy drink, it bubbled and tasted rather like champagne'.

Not only did they praise the taste, but it was also rather healthy. Cook reportedly had no idea that Maori were already using manuka for its health-giving properties, but by using it he inadvertently administered to his crew a great preventative of the disease that struck fear into the mariner — scurvy.

The Wigram Brewing Company in Christchurch produces a version of Cook's beer that is as accurate as they could make it. I bought two bottles. I still have one.

They are refreshingly honest about it, as their brewmaster says: 'There are four million people in New Zealand, and when we sell four million and one bottles, we'll know someone has had one twice.'

It's not a great beer, but if I had been on a boat for 117 days without any beer at all, no doubt it would taste fantastic. And under those circumstances, I might have had a second.

A few weeks have elapsed since the Spring Fling, and Rudolph, my new home-brewing friend, comes around to help me dispose of some surplus apples. There are only so many apples you can eat, so we're going to turn the rest into a refreshing beverage.

It's hard to believe how easy it can be to create alcohol. Creating good alcohol is a different matter though. The ability to turn something into an elixir is not a skill to be sneered at.

We juice the apples, add some sugar, and introduce some yeast to the mixture. Within a few minutes the yeast has burst into life, which is discernible by the bubbling froth that is emerging from the bottle.

Now all we have to do is wait. This is often the most difficult part of the process. That people can wait years, or even decades, for an alcohol to ripen flabbergasts me.

'You know, you can turn anything into alcohol,' says Rudy as we stare at the froth bubbling out of my brew, which I have called applecohol.

'Jeez, Rudy,' I say. 'I reckon you'd be a good guy to know in prison!'

11

Breeding like rabbits

It is really only Shawn and myself (although to be honest, probably not even Shawn) that think my offer to take some problematic rabbits off his hands and transform them into a wholesome dinner is a good idea. Plump, healthy-looking critters with shiny blue-grey coats, they are mighty fine-looking rabbits.

The fact that they have been his daughters' pets may have something to do with his hesitation. But they are now unwanted, as they have grown too big and are too much of a handful for the girls to look after.

I have only mentioned the idea because I heard a radio report that North Korea had imported several pairs of the giant German grey rabbit. They intended to breed these colossal rabbits in an attempt to help solve their food crisis, as a single rabbit could produce around seven kilograms of meat, or as Karl Szmolinsky, the east German pensioner who was behind the scheme, said, enough for a filling meal for eight people.

The plan to alleviate the fate of the starving North Korean masses tragically fell apart after the breeder alleged that the shipment of rabbits, rather than being used as the basis of a breeding programme, had instead been eaten by some of the country's top officials. As terrible as this was, it did lead to a headline in a major newspaper that read 'Godless Commies Ate My Monster Rabbits'. Little wonder the North Koreans are in the Axis of Evil.

Rabbits really are an ideal stock unit for the small landholder.

Considered in the harsh light of practicality, without allowing saccharine images of *Watership Down* to colour our viewpoint, it makes a lot of sense. They breed prolifically (like rabbits, in fact), their meat is lean, and they don't take much to feed. Were it not for their cute little eyes, big floppy ears and the endearing way they nibble their food, they would make ideal livestock.

Farming rabbits for meat is still considered a little repugnant, especially compared to hunting them, but it now appears that I am farming some rabbits of my own, although in a rather more free-range, less hands-on kind of manner.

I had always assumed, probably from watching Bugs Bunny cartoons, that the image of Bugs living directly under the carrot patch so that he could graze at will on the carrots growing through his ceiling had been created with a little artistic licence. But as I have discovered a rabbit hole that has been dug in my garden directly underneath my carrots, the notion is clearly based on fact.

The most astonishing thing about this scenario is that I even have carrots growing in the first place. They have been planted as part of a grand experiment. I am conducting field trials in an actual field.

When I first moved to the farm, I had grand plans for the paddock outside my tent. When looking at its vast grassiness, I'd foreseen rows of potatoes, tomato patches, and tall lines of corn waving gently in the breeze. I imagined gentle paths winding through my lush and bountiful Garden of Eden. I saw wheat and soybeans and pumpkins and watermelons and broccoli and garlic and capsicums and beetroot, and a small rice paddy.

I imagined summer afternoons where lines of sunflowers nodded their colourful heads towards me. I even saw a crude pergola fashioned from manuka, covered in shade cloth, with a picnic table underneath it, a book sitting open where I had laid it down so as to trudge back over to the lemon tree to get another lemon for my drink, past a little windmill that fed the reservoir tanks I had positioned on platforms to allow gravity to feed the irrigation hoses.

How hard could it be to create this simple vision of garden-based Utopia? It isn't like I have a lot else to do with my time.

Of course, big dreams have small beginnings, and Jane decides that what we need is to make a start in the field with a series of three trial beds to test different methods of gardening.

The first is a hay garden. While it is supposed to be constructed with straw, all I have is hay, and I am a little dubious about it. Hay gardens are touted as an ideal no-dig garden, and the theory, as Jane explains, is that you wet the bales of straw or hay, allow them to begin to decompose, and a week or so later plant the seedlings into them. The composting bale will then provide the nutrients needed by the plants.

I throw some bales of hay in the wheelbarrow, and wheel them over to where I have fashioned a boxed border for them. Into the box the bales go, nice and snug, and all I have to do now is lug water over from the trough and give them a good dousing.

In retrospect, I should have built the boxes closer to the trough.

The second garden comprises decomposing sawdust. The sawdust provides the nutrients needed by the plants.

A truck deposits a large mound of sawdust slightly further away from where I need it, but one of the great things about having a television crew in attendance is that they can sometimes get a little bored with standing around, for as surprising as it may seem, there is only so much footage of me wheeling wheelbarrows that Jane requires.

If one is to make best use of the crew in this situation, then I suggest fostering a sense of friendly rivalry. Simply uttering the phrase: 'Wouldn't it be more fun if we made this a competition?' generally does the trick.

Without a moment's hesitation, Frank the soundman and Tom

the cameraman are scrambling to best each other. With Brendan the editor mooching around as well, a vigorous battle quickly develops.

From a standing start, they shovel sawdust into the wheelbarrow, then reverse the barrow back before spinning it on its axis and flinging themselves after it in a madcap dash to where the garden is being constructed, empty it, and tear back.

The competition is fierce and timed runs are close. It all comes down to mastering the key manoeuvres: the reverse and turn, taking the corner by the gate correctly (too tight and they're slowed down; too wide and a few extra steps are added in) and the final sprint for home.

Due to my innate athletic incompetence and natural lethargy, I stand back and egg them on. The more I encourage them, the more they are willing to better their opponent. A quiet 'he said he could take you this time' whispered in an ear, or a muttered 'I thought you might have been quicker that time', and within a few minutes I have three sweaty, puffing workmates, and the sawdust is where I need it.

A quick shot of me finishing the last load and doing a little judicious shovelling, and the sequence is complete.

Sawdust gardens are generally a little on the lean side when it comes to nitrogen, but we manage to track down some bags of used coffee grounds at a nearby café. We liberally sprinkle the grounds onto the soil. I have no idea whether they will work, but they lend the air a delightful aroma of coffee for a while, serving to remind me of the fact that I am not able to visit a café and have one made.

I wonder aloud to Jane if could add coffee plants to my field trial. She says I should concentrate on what I have.

The third garden is more conventional. I intend to plant directly into the paddock. Surely the soil will provide the nutrients needed by the plants?

In order to create this soil garden, I could simply have dug up an area with a shovel, but that seems a little too labour intensive. Instead I hire a small, motorised rotary hoe to till the soil.

The blades of the rotary hoe can barely dig into the hard earth. All the hoe wants to do is to scramble its self-propelled self along the top of the ground at a great rate of knots, dragging me behind it. I have to lean back on the hoe to slow it down, while also forcing the blades down so that they can gain a little purchase.

In this manner, they dig deep into the soil, but then it is a delicate business engaging the hoe's gears and lurching forward, as the blades want to climb back up out of the hole they have dug and run off down the paddock.

All of this is done cocooned in the deafening roar of the hoe's motor. Being tethered like a slave to the smoking, rowdy hoe has none of the nostalgia of ploughing the fields with horses.

As well as the three gardens, I intend to make a start on my grand Eden vision by planting a large crop of corn, which should be easy to grow and its tall stalks will provide an element of privacy from the peering eyes of the neighbours.

If I thought digging the small area for the soil garden was difficult, it is nothing compared to the area needed to be tilled to plant the corn seed. After a few hours and many passes of the hoe up and down the strip, with hands itching from the constant vibration, ears ringing from the noise, and eyes weeping from the smoke, I complete the preparation.

This is not one of those jobs where you look back at the end of the day and see what you have achieved. Instead I look back and say, 'Is that all I've done?'

Had I planted the seeds there and then, I would have been fine. The soil, freshly hoed, is soft and pliable. However, I suggest that we should probably graze the area off first with the landlord's heifers.

I should probably have done this before I constructed the gardens.

The cattle find the tilled soil fascinating, and all they really want to do is stand on it. When they tire of standing on it, they dance little high-kicking cattle jigs on it.

When they encounter the garden with the coffee grounds sprinkled over it, they go crazy, licking the soil, and prancing around dementedly. Within a short period they have done a wonderful job of compacting all the soil back down.

Then it rains. Rather than softening the soil, it hardens it even more, so that when the time comes to plant the thousand or so corn seeds, I can't even push my finger into the ground the required depth.

Using the steel handle of the potbelly stove door as a probe, I prod repeatedly into the hard soil to make holes for the seeds. Because I am using the palm of my hand for extra leverage, I soon wear the skin off the centre of it. Kneeling on bruised knees, I swap hands, and within a few moments wear the skin off the other palm. My bleeding palms make it appear as if I have developed stigmata. My garden is more a Golgotha than an Eden.

Even if I only get a 10 per cent strike rate, I will still have over a hundred corn plants. What could possibly go wrong?

Pretty much everything, it seems.

The main problem with having the gardens in the middle of the paddock is that the rest of the paddock needs to be grazed by my livestock. Simply asking them to refrain from eating my produce won't work, so what is needed is a fence.

The electric fence is great for those who don't have the resources or effort required to build a decent fence. I suggest to Jane that I could run an electrified wire straight from the electric fence unit in the shed to a fence around my garden.

While that will most likely deter the stock, the drawback is that there will be a fence right around my gardens producing a massive electrical output. Of course, as I will no doubt do a half-arsed job, I will have to spend a considerable amount of time and effort avoiding shocking myself.

One of the first mentions of the electric fence occurs in Mark Twain's 1889 novel *A Connecticut Yankee in King Arthur's Court*. However, it wasn't a Connecticut Yankee but a Waikato farmer who revolutionised the electric fence.

I'm not sure whether Twain's novel inspired Bill Gallagher's decision to electrify his car. Bill stated that it was his wish to stop his horse Joe from rubbing himself against the vehicle (not his desire to stop Mrs Gallagher leaving the farm on shopping expeditions) that led to his combining Kiwi ingenuity with some Meccano to electrify the car. Reports say that Joe soon stopped his scratching, but they are strangely silent on the fate of Mrs Gallagher's shopping expeditions.

Bill was nothing if not practical. After realising that electricity was a wonderful way to control animals, he promptly attached a mile of his fence directly to the mains.

He was no doubt disappointed when informed that not only was it illegal to attach a fence directly to the mains, but that it may also not have been the most sensible of ideas. Undeterred, he invented a battery-powered system, which launched a successful company.

The barn at the farm has an iconic orange Gallagher electric fence unit installed, where it click-clicks away rhythmically until we think it wise to turn it off, as no one wants an unexpected belt from an electric fence.

Attempting to work with wire that is electrified is an excellent exercise in dexterity and mental control. Reaching through live wires to attempt to untie or tie them is a lot more exciting than the likes of the game Operation. The annoying buzz that alerts you to your failure to remove the Water on the Knee is nothing compared to the body-jarring judder that one gets from the electric fence. That can result in Water on the Floor.

My grandfather used to love grabbing electric fences. In many ways I think it is the reason he has lived into his nineties. Every so often he would shock some more life into himself.

I think of him as I tentatively try to see if there is current in the fence by testing it with a blade of grass. Grass is a poor conductor of electricity, and by applying it to the wire it is possible to feel whether the fence is electrified.

I'd already had a bit of an electric fence failure on the farm. In the first few days I'd strung a portable electric fence across the paddock to keep the cows away from the tent, but then Shawn arrived, took one look at the fence, and asked me if I had checked to see if it was live. I said I had, using the aforementioned grass method, and that it was.

He then pulled out his flash little current checker, and checking the fence in several spots quickly ascertained it was not.

I had only tested it at the end where I had connected it to the other fence, assuming that the current would then traverse the length of the fence. It hadn't. The problem, it seemed, was modernity.

When I was young, the portable electric fence was rugged and heavy and made from wire. Nowadays, the electric fence is lighter and made from some kind of plastic strapping with strands of thin wire running through it to conduct the current. It is an effective unit, but only when the current is turned on.

When the current is not on, the fence becomes a tasty chewing treat for cattle. Their mastication breaks the wires, rendering the fence effectively useless as a transmitter of current.

It struck me that here is a pertinent symbol of the fallibility and unsustainability of the modern world.

Jane won't allow me to use the electric fence to protect my paddock gardens, as there is a more sustainable solution, which, unsurprisingly, entails a great deal more effort on my behalf.

Lying in rolls of various sizes around the property are lengths of wire netting. These will be ideal to keep out prying livestock, Jane says.

As I do not wish to dig in posts, I use a combination of warratahs (steel Y-shaped poles that are hammered into the ground) and bamboo poles that I have brought from home, as I have a large bamboo hedge. Anyone who has a bamboo hedge knows just how infuriatingly sustainable a crop it is. It's virtually impossible to eradicate. Thank goodness I have finally found it useful for something.

Within a short time, I have constructed a fence around my crops, and I stand back to survey my handiwork. Harriet and Coco wander over to survey it too.

Coco likes it so much that she approaches the fence, sniffs it, and decides it will also require viewing from the other side.

I can't say she walks over the top of the fence exactly — it's more like she actually walks through it. I have no idea how. It's as if the fence does not exist.

Over the next few months I attempt to make the fence more cowproof, but it never works. Jane will use its failure as a constant source of ridicule, which seems unfair as she has so many other sources to choose from.

In the end it doesn't really matter that the cows can get into the gardens, as there isn't really much for them to eat there. The grass may be greener on the other side of the fence, but my vegetables are not. Coco does eat some of the hay from the hay garden, so perhaps it isn't entirely wasted.

If the fence is inadequate at stopping cows, it is even less adequate at stopping rabbits.

I pull some of the small but perfectly formed carrots up from above where the rabbit has burrowed underneath them. They are untouched. The rabbit has gone to all the effort of digging a burrow under them, only to be disappointed by what it finds.

I can well understand its frustration.

12

Pansy acts the goat

I have suffered a devastating loss. My sheep have run away. Or they may have been rustled. I prefer the latter option, as I would hate to think that I have so traumatised them that they have upped and left of their own accord. I feel violated.

I had to go away to work on another project overnight, and returned to find only empty paddocks. There is some debate about how they escaped. There had been several people at the farm the day before, so whether someone hadn't closed the gate properly, or whether someone had driven in and taken them will never be known.

It could have been worse though, as Harriet and Coco were initially missing too. Thankfully, I find them a short distance away down a side road, grazing peacefully.

Now there is just the question of how I will get them back. I am fairly sure that where Harriet goes, Coco will follow, so I decide to catch Harriet by the halter and lead her back. Only Harriet doesn't seem to want to be caught by the halter.

Eventually I find a bucket and put some stones in it, thinking that she will be under the impression they are cow nuts. The deception is enough to lure Harriet in, and grasping her halter I lead her back up the road, with Coco reluctantly tagging along. I can only hope that a car doesn't spook her. That would be a disaster for everyone involved, especially Coco.

With the cows safely returned, I drive up and down the road, looking into driveways, through fences, into paddocks, and in people's sections in a vain attempt to find the sheep.

To the untrained eye, one sheep may look a lot like another, but I am in the fortunate position of having just shorn mine. They are fairly distinctive, and I am pretty sure I will recognise them in a mob, even if they are in a mob of other freshly shorn sheep.

I even go so far as to put up some flyers with their picture on them. It isn't the most flattering picture of them, taken as it was post shearing, but now is not the time for sheep vanity.

As the days pass, the general consensus is that the sheep are gracing the barbecues of the rustlers.

While I was never emotionally attached to them, I do miss them, as they had been doing a good job of keeping the grass under control around my gardens.

I have foresworn mowing, so having the sheep and cattle was a pleasure. When the grass was quite long, I had leased out the paddock as grazing for my landlord's stock for a few days. I sat in my thinking chair as they mowed the grass off with their long tongues. But they couldn't be trusted to maintain the grass inside the enclosure of my field trial gardens without plundering what little produce was growing there.

While I had been making cheese with Sabine at the Otamatea Eco Village, during the hour the cheese did its cheesy thing, we strolled down to the orchard where her husband Wolfgang was mowing his meadow with a scythe.

Wolfgang is over two metres tall and is very lean. With his angular face and armed with his scythe, he is the personification of Death as depicted by medieval drawings and heavy metal album covers. It all looks so Germanically appropriate.

I have only ever seen a scythe used in old film footage, yet here he is wielding one with a fluid and graceful motion. I am flabbergasted at the ease with which it fells the foot-high grass.

'It's meditation in motion,' Wolfgang says.

He offers me a go and it's a simple task to master running the blade of the scythe parallel along the ground, with the rear of it just resting on the damp earth, so that the blade effortlessly hews the grass.

It is razor sharp, and Wolfgang keeps a whetstone in a bamboo water carrier at his waist to keep it that way. It isn't used to sharpen the blade as such, but more to reshape the wafer-thin cutting edge so that it stays straight and true.

The scythe is a remarkable piece of technology, and using it is a wonderful, rhythmic, silent way to mow the long grass.

It also gives my disco muscles a good workout, as most of the movement seems to come from the hips.

Wolfgang explains that in Germany there is a measure of land known as a 'morning', which is the area a man could be expected to be able to mow with a scythe in that period of time. It's about a quarter of an acre. I don't know if there is an area known as an 'afternoon', or whether the efforts of the morning mean that an 'afternoon' is the designated length of time a man could spend at the tavern before he has to get home for his tea.

'Why don't we see people using these any more, Wolfgang?' I ask.

'I think the skill is lost; farms got bigger, oil was cheap. That's about to come to an end, of course,' he says. 'But it's much quieter than a lawnmower. There are no fumes, and it isn't really that much slower.'

It is time to bring the art of scything back, I think, as I survey the grass growing inside my garden plot. It has become quite long since the disappearance of the sheep, and is causing significant dampness to my lower pants area.

In lieu of my handy sheep, I have arranged a challenge between Wolfgang and myself to see who can mow a meadow more efficiently. He has graciously accepted.

On the appointed morning, Wolfgang arrives. Regrettably for

my overblown sense of drama, it isn't sickles at dawn as such, as he arrives about nine. Gentlemen's hours.

It is Wolfgang and his scythe versus me and my weed-whacker versus Pansy, a brown-and-white goat, and her stomach.

I have borrowed Pansy for a while to see how she fares against the two of us. She is proving to be a charming animal.

As a child we had an old billy goat who was supposed to keep the grass chewed down around the place. His name was Billy and he was ferocious. The only person who could deal with him if he came off his chain was my grandfather.

At my grandmother's funeral, my uncle recounted the traumas that Billy had inflicted upon her. She'd had wonderful gardens, as countrywomen of her era did, and when Billy slipped his chain he would habitually make his way there to graze on her flowers.

Once, when Grandma was kneeling down weeding the garden, Billy had snuck up behind her. When she became aware of his presence she fled, pursued by the malevolent goat. She managed to get inside the house, but Billy followed her in, and she was forced to barricade herself in the bedroom while the goat sauntered around the house.

Grandma spent a terrorised few hours trapped while the goat toured her immaculate home. Eventually Grandad came home and rescued her from her horned nemesis. I imagine he didn't dare have the courage to laugh.

Pansy couldn't be more different to Billy.

While we often see goats kept alone, they are very much a herd animal, and Pansy hates being tied up by herself. I could let her roam around, but her intelligent and inquisitive nature invariably leads her to eat everything she isn't supposed to, and to want to investigate everything, including the inside of my caravan.

She is particularly adept at getting through fences. Of course, having horns can make reaching through fences to graze problematic, as they go through all too easily, but get stuck on the way back out. Many will be the time during the project that I will discover Pansy, kneeling down, head trapped in a fence by her horns, looking dejected and a little shamefaced. This propensity to wander means that she requires constant supervision. Without a doubt, she is the most labour-intensive animal on the farm.

Naturally, Pansy won't eradicate the grass as quickly as either Wolfgang or me, but she seems an apt competitor, albeit the one most likely to veer away from the target area and give my valuable gardens a thorough mastication.

I rig up some set areas of equal size and difficulty, and mark them with fencing tape and barrels, and off we go.

By the time I have mixed the fuel for the weed-whacker, Wolfgang is well under way. Obviously, I could have mixed the fuel sooner, but that seems unfair, as in my experience no weed-whacker ever has fuel in it, or its whacking string on it when one wants to use it. Therefore this must be added to the duration of the process.

By the time I wind the plastic string on, Wolfgang is almost half-finished.

Pansy appears to be transfixed by the scythe, and isn't doing much except watching Wolfgang.

There is a cold bead of dread about my spinal area as I contemplate the fact that I have yet to start the infernal machine, and the notoriously unreliable two-stroke motor has over the years driven people to madness.

I bend down, thumb the ON switch on, set the choke to what I believe is full (I have never really been able to tell as there are no markings on the machine), pump the plastic nubbin of the petrol primer vigorously, and pull the cord.

Miraculously it fires into life, smokes, coughs, and continues to run as I reduce the choke to half, massage the throttle, and burst into that spontaneous smile one gets when a two-stroke motor starts at the first pull. I'm off.

Wolfgang refuels by taking a leisurely drink, watching as I tangle the cord in some wire.

Pansy stares at us both while nibbling at a dandelion.

The shrill shriek of the motor rises as I apply the throttle, prompting an intoxicating cloud of oily smoke to be expelled abruptly from the exhaust. The motor strains as the spinning plastic whacker string shreds the soft, yielding grass. It doesn't cut it so much as tear it bodily apart, leaving tattered stumps of grass where the scythe leaves razor-like cuts.

The weed-whacker flings the cut grass all over the show, whereas the grass Wolfgang has cut lies in neat furrows. It takes him little time to collect the grass he has cut, whereas I still face an eternity of raking.

Wolfgang finishes, having barely broken a sweat, and is standing chatting to Jane.

I am barely halfway through.

Pansy has tangled her chain around her pole, and tired of our antics and unable to go anywhere, she has sat down for a nap.

Wolfgang sums it up best when he says, 'We think of it as work, whereas she thinks of it as just living.'

It's the last time I will use the weed-whacker. I don't give it up for any reasons of sustainability. Instead, I have somehow done something that causes little shards of metal to fall out of the motor.

After having started so easily, it never goes again.

13

A sow's ear

I have never felt so much like a wildlife filmmaker as I do standing in the middle of the forestry road, holding the tracking device that will locate the wild pig.

Well, technically it will locate the dog that has located the pig, and it will locate the dog even if the dog hasn't located the pig, but it is nonetheless a wonderful moment in my television career.

The situation is emotionally enhanced by the knowledge that when we do locate the pig, I will then stab it to death with a knife.

We can hear the dogs, barking and crashing through the scrub, but then all goes suddenly quiet. The air is so tense I could practise my pig-stabbing technique by cutting it with a knife.

It is with a heightened sense of awareness and a grim sense of theatre that I allow the radio receiver to drift in a repetitive arc to see if I can locate the exact direction of the dog. The T-shaped receiver emits an intermittent buzzing sound, which rises to a single note when pointed in the general direction of the unseen dog.

Along the track my hunting companions stand quietly alert, gazing off into the scrub, listening attentively to see if they can hear what is happening.

I'm with Glen Osborne, former All Black and general wag, his buddy Matt the butcher, and Matt's young sons. Having the boys present highlights the intergenerational nature of the hunt. The closest most children will ever get to a hunt is reaching into the meat chiller at the supermarket to buy a little disembodied dinner.

It never ceases to amaze me that we have vast tracts of bush where it is possible for ordinary citizens of any means to go hunting, and to do so for free. This is not the case the world over. It would have been a source of immense frustration to have lived as a 19th-century peasant in the forests of the local lord, and been forbidden from hunting any of the game you encountered, even if your family was starving.

I always experience a sense of nostalgic nationalism when, driving around the provinces, I pass a 4WD, usually with dog boxes on its deck, parked by the side of the road in a forest. Somewhere in the surrounding area there's a primal struggle of life and death being enacted. A story being told that is as old as man — the hunt.

The feral pig is a delightfully exhilarating target species. The very fact that they are described as feral is in itself superbly evocative, conjuring up images of hairy low-slung hogs, stout of shoulder, tusks dripping with slather, turning to confront their aggressors. Today that aggressor will be me.

Even the tools that are used to chase them, namely dogs and knives, remain largely unchanged from the earliest days of hunting.

The pig dog is a composite mongrel bred to hunt, doing as nature, and a selective breeding programme, would have them do. Big dogs, small dogs, finders and holders, they bound from the back of the ute, all bright alert eyes, tightly sprung muscles, and barely contained excitement.

It is an undeniably primal experience.

While some things remain the same, there is a touch of modernity about our proceedings. Not only do we have the tracking devices, but Glen carries a short-barrelled lever-action rifle. It's a beautiful weapon — stocky, and perfectly suited to the bush. It would have made an ideal weapon in the New Zealand wars, serving as both rifle and club.

Glen is also wearing leather chaps. They result in a round of good-natured banter as to the lifestyle choices that leather chaps represent. As much as we chastise him, I am a little envious. They appear practical, as well as stylish. I only have on a light pair of jeans, and not being a big fan of gorse, I am not looking forward to forcing my way through it in a frantic manner.

Gorse is needlessly needle sharp, and tends to impale you wherever possible. Even when dead it's a prickly curse, easily penetrating light footwear. If the people who imported it here to serve as nicely manicured hedges aren't being flogged with strands of it in one of the many levels of hell, then there is no justice.

My jeans are held up with a piece of baling twine, onto which a sheathed knife had been added. According to Glen and Matt, I have not yet earned the right to wear a belt, because I have not yet been blooded.

It's a colourful term, blooded. It has a certain romantic barbarism to it. I like the symbolism of it in a society that doesn't seem to have a lot of symbolism.

Matt's young boys already have their belts. Compared to me, they are seasoned, albeit decidedly ungrizzled, veterans.

The occasional bark echoes around the clearing as we loiter, a tight and tense little group, waiting for a signal that there is contact. Conversation is muted, eyes scan the gorse. I wave the finder around. The camera crew film me doing it.

A dog barks, then another, then there's the sound of brush crashing down, and more barking. Almost before we can move, the squeal of a pig joins the dogs in a chorus of savage intensity. Headlong into the manuka and gorse we plunge.

A change in the tone of the dogs' barking is drowned out by the louder screams of the pig, as scattered shouts of 'This way' and 'They're over here' and 'C'mon, Radar' sound from various directions.

I blunder through the gorse in as manly a fashion as my timidity allows, all the while knowing that the situation into which I am running is going to be one in which I will encounter one of the forest's largest and most dangerous animals, all tusks and diseases and lice.

Actually, it's about the forest's only dangerous animal. It always seems as if more people are killed and hurt by the forest itself than by anything in it.

Running just behind Matt means that I catch the branches that he's pushing past. They whip back at me, catching me on my forearms, which I am holding up to protect my beautiful face. It's not Matt's fault — there's no time for niceties. Who knows what we will encounter? Will the pig have turned the tables on a dog, eviscerating it with a tusk?

The barking is louder now. I certainly don't need the radio receiver any more. It's fairly easy to locate them — they'll be at the place from which an unending and unnerving squeal is issuing.

Suddenly I'm there, and there it is in the low manuka regrowth. A black, hairy mass of muscle is trying to turn and fight off each of the dogs in turn, but as soon as it goes for one, the others bound back in. It's clearly a musketeer-like world for man's best friend. All for one and all that.

The dogs are now clamped onto various parts of the pig's anatomy, pulling it out in several directions, effectively pinning it, while it screams a squeal of penetrating angered anguish.

The noise alone is a good incentive to kill it immediately, to spare its suffering and ours.

Squealing is the pig's forte. Picking up baby pigs will elicit a squeal as if you are torturing them, then when put down they waddle off as if nothing ever happened. This time, though, something is happening.

'We got it, Radar,' says Glen in a masterful moment of understatement.

'Great, what do we do now?'

'Now,' he says, 'you're going to kill it.'

Oh. Right.

I have the knife. The pig, one of its ears torn off, is still squealing. With the dogs still harassing it, we grab the back legs and then flip it over. It's kicking and thrashing, but Glen and Matt work together with a precision that says they have done this a number of times.

Grasping the stout front legs, they splay them open, presenting its throat. Matt indicates where I need to create an incision in the throat that will allow me to plunge the knife into the beast's heart.

Opening the throat is no problem. I use the sharp point of the knife and slice a short straight slit. I'm acutely aware of the combination of fearsome tusks and the snarling dogs not far from my hand.

'The knife's not long enough,' Matt says. 'It's a bit bigger than we thought. You'll have to stick your hand right inside to get at the heart.'

I now have the knife and my hand entirely inside the throat of a live and rightly aggrieved pig. For a split second, my brain thinks to itself that on a chilly morning such as it is, the sensation is one of pleasant warmth on my cold fingers.

I make a couple of brisk futile thrusts before I feel the knife connect with something that seems to have the solidity of a heart, and I plunge the blade into the mighty organ, give it a mighty wiggle, and unleash a torrent of black blood that spews out of the gash in its neck, over my wrist and onto my pants.

The pig struggles, becomes weaker, thankfully stops squealing, and is still. There's a moment of calm, almost of tenderness, as we release our hold on the pig and haul the dogs off.

Crikey.

Blood oozes limply over my wrist as I withdraw the knife. 'That's a big pig you got yourself there, Radar,' says Glen.

It certainly is. Wide of shoulder with chunky thighs. I stare at the thick bridge of its nose. An ear is missing, ripped off by a dog in the frenzy of the attack, leaving a slick red welt.

The ear offers the dog an easy place to hold the pig, and it is said that some hunters will cut the ears off the smaller pigs or the breeding sows they catch, making it much more difficult for the dogs to get a purchase.

I turn to Jane. 'I suspect we won't see that bit on the telly,' I say.

'No,' she agrees. 'We most certainly will not.' Only this time, if you watch the episode, we do, albeit fleetingly. It's hard to hide the brutality of the event.

I open the pig's belly, revealing a mass of fresh, glossy innards, and we check the liver to make sure it isn't blighted with anything untoward.

I set the liver and kidneys aside, and we haul the rest of the offal out onto the ground, leaving it where the pig fell, no doubt to be eaten by other pigs not affected by the kinds of morality we humans are.

It's quite something to watch the delicacy required when cutting out the rectum of an animal, so that everything can be removed without spilling anything unpleasant.

The best wild pork I ever had the pleasure of was a leg I brought back from the Chatham Islands. It weighed several kilograms, took four hours to cook, and had that darkness that symbolises the flesh of the wild pig.

I also brought some weka back with me, for in the Chathams the weka is an introduced pest and able to be hunted. If you are ever there, I recommend asking a local for one. They taste slightly less fishy than a muttonbird.

The pig is a lot heavier than I imagined it would be. It's a dense, dead, floppy weight. It's almost on the cusp of being too heavy for little old me.

We tie the legs together so that I can wear it on my back like a pack, but Glen, because he is Glen, is tying the knots looser than they should be. This results in the knots coming undone and the sheer ungainly mass of the pig sliding off my shoulder and dragging me down with it.

I know he is doing this because he is a prankster of some repute. Making mirth is not often bettered when the foil is the carcass of a dead swine.

Each time I fall, they retie the knots, grab me by the hands and help haul me up into a swaggeringly erect position. I adjust the weight of the hog, and lurch off into the thicket. The rough ground and uneven track cause the weight of the pig to shift, the rope to come undone again on the legs, the pig to fall off and drag me down on top of it, with the result being that I flail around like some kind of dementedly upturned man-beetle.

It's a brutal ritual, the harvesting of the hog.

We need to heat a bath to heat water to scald the pig, and then shave it. I'm handed a pair of tall gumboots, and when the water is hot enough the pig is plunged in. I get in and stand on top of it, kicking the hair off as the heat releases it, making sure the pig is well submerged and becoming hairless and slightly translucent.

The rest of the shaving is done with the knives, sliding them across its water-lubricated belly, removing the hair in a peculiarly touching moment of rough familiarity. It's bizarrely intimate.

I have ended up blooded. There's no fancy gesture needed — I am literally soaked in blood from head to toe. The back of my jersey is wet with blood. It's in my hair, it saturates my jeans, and stains my hands.

At the end of the day as I hand Matt back his knife he says, 'Nah, you keep that.'

One of the sons pipes up and says, 'But Dad, isn't that your good kni—' before Matt shushes him.

I'm deeply touched.

After the intensity of the morning's undertaking, I toy with the idea of stopping off and splashing out a little of my stipend on a can of a popular carbonated beverage at a dairy. I choose not to, and in retrospect realise this is a wise idea, as I'm sure the bloodstained clothing and the knife may take some explaining, especially as I am not the picture of the stereotypical pig hunter.

What probably isn't such a wise idea is deciding that I will wash the blood off by taking a swim in the sun-dappled sea, in an area not far from where shark sightings have recently occurred.

14

Busy little bees

I would like to think that if I am pinned by the lower arm or leg by a boulder, or a tree, or an overturned tractor, that I'd have a fair idea where I would cut through the limb in order to extricate myself.

I'd go straight for the joint, at either the elbow or the knee. Surely, I tell myself, it's better to cut through skin, flesh and ligaments, before finally snapping the limb off at the joint, than to attempt to cut through the main bones. Who has the fortitude for that?

If I were trapped higher than the elbow, then I'm sure cutting through the shoulder joint shouldn't be too much of a problem, but quite what would happen if I had to cut through the thigh I don't know.

Regardless, it would sting, and the whole process is no doubt easier thought about than executed.

Then there is also the problem as to whether an amputation at the joint makes it more difficult to fit a prosthesis.

My strategies also assume that I would have access to a knife or saw, and while I do have a knife, I have yet to buy a sheath for it, so I don't carry it around. I am clearly not as prepared as I could be, despite my Scouting background.

I always pay careful attention to how people survive these incidents, because there are any number of things that can go wrong when you're stranded alone in the semi-wilderness of a paddock conducting a sustainability experiment.

I am acutely aware of many of them, thanks to a childhood crammed with safety warnings. As children at a rural primary school, we were often taken on special field trips to watch gruesome displays and be shown some of the most graphic photographs of injuries I can ever recall. We were shown photos of all the myriad forms of mutilations caused by machinery — medical snuff photos — to remind us that what we encountered on a daily basis were not toys.

We were shown deep gouges in people's heads, shredded limbs, crushed hands and lacerated legs. We saw cloth mannequins mulched in hay balers or flayed by PTOs.

While to fresh-faced town children a PTO might be a polite request to Please Turn Over a storybook's charming page, to a rural child the PTO is the Power Take Off, a metal driveshaft that emerges from the rear of a tractor to power the attached implements. It spins at over 1000 revolutions per minute, and is notoriously unforgiving to those who have the misfortune of being caught in it courtesy of any loose clothing or carelessly placed limb.

As children we were reminded that it was not only the high-powered driveshaft that was dangerous, but also the implements it powered: the heavy spinning blades of mowers and rotary hoes, the hay-turners' steel tines, and the chain drives of feed-out wagons. There was all manner of unrelenting steel waiting to slice, crush, or ensnare the unwary.

Of course, the manufacturer's specifications said that these implements were not to be approached while engaged, but the practicality of farming overrode this on most occasions.

I was brought up by my parents to have a healthy respect for this kind of machinery, which I was allowed to clamber over, and even operate at a young age. I'm sure that accidents may have been closer than I realised. But I survived. Not everyone did.

Now I am back on a farm, and everywhere I look there is danger.

As an adult fully apprised of the inconvenience of mortality, I am forced to confront my childhood apprehensions and farm-based trepidations.

It hasn't helped that I have read too many *Reader's Digest* Drama in Real Life stories, where people caught in combine harvesters have to crawl 30 kilometres with their legs severed, or fight off a bear that is intent on eating their face.

I could cut myself, perhaps on some tin, maybe with a knife, hopefully not with a circular saw. I could hit my thumb with a hammer, and the nail could turn black and fall off. More than likely I will burn myself, catch a cold, or walk into something. I could well burn myself after walking into something that cuts me.

There are any number of plants that could give me a nasty rash.

Splinters annoy me.

There's always the danger of a piece of wire snapping back to impale me in the eye. Unfortunate cowpokes in Westerns were frequently entangled in coils of barbed wire that had snapped as they were tightening it. These men were usually alone on some desolate part of the range.

That's the thing about farming accidents: they will most likely happen while you are alone and some distance from help.

And the hygiene is not so great either.

I often wipe my hands and cutlery on the grass in order to clean them. Who knows what dangers lurk there? Scientists know, I guess. Or at least scientists who specialise in the field of fields.

And then there is the problem of zoonoses. These are diseases that can be transmitted from animals to me. There's leptospirosis, brucellosis and tuberculosis. The common factor: the cow.

Don't chickens have salmonella? They drink from the water trough that I occasionally bathe in. What else could be in the trough? When I'm soaking in it, there are lots of little things that bite me. What are they? Could they burrow into my skin or, heaven forbid, enter me through a much more private and embarrassing route?

And what about the algal slime that grows in the trough? While it is more than likely harmless, what if I were to slip on it and fall as I'm getting in or out of the water?

If animals don't give me diseases, there is always the danger of being trampled. Milking the cow means there is the risk of being badly kicked. A sheep could butt my knee. At least there is only a limited chance of suffering fly-strike.

Let's not forget the chance of sunburn.

Then there is the danger of the compost pile, which can contain spores. I don't want to be incapacitated by a spore. Nor do I want to contract a flesh-eating disease.

I could poison myself by eating the wrong part of a plant. I could fall in a hole. I could tweak a muscle attempting to do something that is physically beyond me, which is most things. A hernia is a possibility. Haemorrhoids are, no doubt, a given.

Bales of hay could fall on top of me.

Perhaps worst of all is the knowledge that I could be killed by a bee.

I recall as a child one day encouraging my parents to come around to the side of the house to look at monarch butterfly chrysalises, when a pain in my gumboot distracted me. A bee had rather stupidly flown into my boot's murky depths, and had promptly stung my leg. Within minutes I was itching and swelling. My breathing became laboured. This no doubt caused concern for my parents, but I don't recall being overly traumatised.

A very quick trip in the Rambler Rebel into Huntly, a shot of adrenaline at the doctors, and I survived. Obviously.

After that I had to carry pills with me in a stainless steel amulet.

I have been stung since, and I haven't always had to take the pills, but I guess the possibility still exists. In any case, it's an excuse to wear an amulet — or it would be if I could remember where I put it.

The last time I was stung by an insect, I was in the middle of the

Sahara shooting an episode of *Intrepid Journeys*. I felt a stinging in the knee area. I put it down to one of the small prickly seeds that littered the desert, like a spiky biddybid. But the stinging moved, and I shook my pants. Onto the desert sand fell a scorpion. I can't say that my life flashed before my eyes, but thoughts of my imminent death certainly did.

To compound the problem, I was filming a piece to camera at the time and I didn't want to be seen to overreact to what at that stage I was convinced would be at the very least a near-death experience.

'Umm, I've just been stung by a scorpion,' I said.

I began to wonder how long it would be before there was a reaction to the toxin. I did have with me a vial of liquid adrenaline in case of a bee attack, but quite where it was I wasn't sure. Would the toxins be related?

What a glamorous way to go: on assignment in the middle of Africa.

Gosh, I really didn't want to seem like a wally and blub at the thought of my demise. Stiff upper lip and all that.

Actually, were my lips stiffening? Maybe. I wasn't sure I could feel them; but then, could I feel them normally? Was my breathing becoming laboured? Or was my preoccupation with my breathing causing my breathing to be irregular?

Was this a panic attack, or was I going into anaphylactic shock, which would result in my body shutting down?

Would they take my body home? I'd smell pretty ripe in a short period, unless they stowed me in the cool store with the food and beer as the festival carried on around me.

We rushed off to find help, and ran into the shady guy who was running the show.

I had squashed the scorpion and put it in a tissue so that they could identify it and administer the correct antivenom. What if I should die for a lack of antivenom? How long would I have? We were 80 kilometres north of Timbuktu, not a great distance, but still a distance that meant several hours of travel over an unforgiving desert landscape.

He took one look at the scorpion and said, 'It's only a baby, not at all dangerous. Would you like an aspirin?'

I settled for a beer.

The year 1979 saw both the introduction in New Zealand of the car-less day, and the discovery of the Asian paper wasp on our fair shores. While the car-less day disappeared within a year, the plague of the paper wasp has become an ongoing problem, as Frank the soundman experiences, in a most delicate place.

A wasp has stung him in the groinal region, and he has decided to take matters into his own hands and eradicate the nest of them we have discovered in a hole in the paddock.

Now I am worried that the fire brigade will be summoned as Frank's eradication method has gone beyond the liberal amounts of fly spray we have been using, and now includes a smoky combination of methylated spirits, a car tyre, and matches. The plume of smoke is rising into the still summer afternoon.

Wasps are flying around, looking a little disturbed at the incineration of their house. Goodness knows how many of them are down there, but the hole verges on huge.

For weeks we have been eradicating nests of paper wasps with fly spray. I know that it isn't overly sustainable, and no doubt there is a fair amount of collateral damage, as my method is to send large clouds of it drifting over the area of the nest to quell those flying about, before making a quick foray to the nest site and coating it with spray until it is a foamy white mess. I don't so much spray them as drown them in spray.

While I realise that they are a part of nature's ecosystem, they also present a danger to us all as they have decided to try to colonise areas we frequent, such as the woodpile, the piles of usable trash, and on fences. I do not want them to attack Jane's child, or me.

Quite often it is difficult to locate the nest, so it requires much wasp watching to see if it is possible to discern where they are centred. They fly along with their legs hanging beneath, a dangling undercarriage that makes them look a little comical, and slightly lazy. But they are busy little creatures, and much of their business seems to involve collecting fibre that they bind together with saliva to form a honeycomb-like nest, into which they deposit their brood. Their other business is concerned with stinging us.

I'd like to rail against the blight on our country that is the introduction of foreign species, but to be fair, we can't claim the moral high ground. This high ground is occupied by *Artioposthia triangulata*, or the Canterbury flatworm. Yes, one of our most successful livestock exports (after the 1924 Invincibles) is the Canterbury flatworm, which in the 1960s succeeded in colonising vast swaths of the United Kingdom, where it proceeded to slaughter on a wholesale level the indigenous earthworms by anaesthetising them and reducing their innards to the consistency of a soupy gruel, which they then consumed for sustenance.

The official advice from the British junior agriculture minister was to eradicate the Canterbury flatworm by treading on them, which prompted a question from the House of Lords that asked: 'Are you sure it's effective to stamp on a worm that's already flat?'

15

The right bait for whitebait?

As the light drizzle turns to an inclement rain, I struggle through the long grass and tall reeds, armed only with a fry pan, my camp-burner stove, five eggs, and a heady sense of optimism. With the reeds and grass towering well above my head, the entire place has the feel of a jungle scene from any number of Vietnam War movies. Not a bird sings. It's wet and eerie.

I have to confess that I am a little worried that at any second I may be the victim of an ambush by either a flurry of black-pyjama-clad Vietnamese or a surge of trigger-happy American GIs, either of which will result in me being killed to death.

I hadn't realised that it is possible to catch whitebait in a creek off the Kaipara Harbour. Apparently it is, or at least that's what I've been told.

They are a funny fish, the whitebait. They seem to be mostly eyes. The thought that we eat them whole — eyes, organs, their stomach full of whatever it is they eat, and even their anus — never seems to faze people. Perhaps they simply don't think about it.

If you are a whitebait fan and you've never thought about it until this point in your life, then sorry. You'll think about it next time, I can assure you.

So valued is the tiny fish that wars have been fought over them, or if not actual wars, then certainly spirited skirmishes, or at least feuds. I suspect there is nothing like a good feud to make a meal of the delicate little fish taste all the better.

Even though I have never been whitebaiting before, there's a sense of nostalgia associated with it. They are a hallowed New Zealand food-based institution — although I suspect that many of us have eaten them with no real idea what exactly it is that they grow into.

Scotty, a local fishing enthusiast, has promised to help me catch 'a feed' of the wriggling delicacy. 'A feed' is such a wonderful New Zealandism. In its ambiguity it also neatly sums up the concept of the sustainable harvest.

Crashing through a stand of dripping ponga, clambering over a fallen log, and falling through a clump of flax, I look up and there is Scotty. He is crouched beside the rain-splattered creek next to his net, the top of which is barely protruding above the water. It seems a relatively random place to have thrown the net in, but I assume he knows what he's doing. Personally, I have no idea.

Scotty is a slightly rotund chap, with a beaming, faintly lop-sided smile, and is cloaked in an air of stale rum. He's a little the worse for wear after having spent the night with a friend who has recently returned from abroad with the obligatory duty-free booze. Still, what better way to overcome a hangover than to sit in the drizzle by a creek, waiting for the vagaries of the whitebait to play out.

While the conditions may not be ideal for whitebaiting, they are certainly less than ideal for filming. Tom the cameraman has trouble with the warm rain, which continually fogs up his view-finder, and as there are times he cannot see anything, he can only point and hope. The result is what appears to be a succession of artfully composed shots of sticks in the foreground of the action. It looks less like we are being filmed, and more like someone is voyeuristically observing us from the safety of the bush.

It's been raining all day. It rained yesterday, and the day before that, and the day before that, and the day before that. While the day before that had been fine, the day before that had not. It has been raining for so many days that each night before I go to sleep I check to see if any webbing is developing between my toes. While there does appear to be some kind of something going on with the skin between my toes, so far it appears less like the much-desired webbing and more like a fungal infection.

The weather is one of the things that people who don't work outdoors tend not to experience in all of its varying magnificence. For most people, rain is something to be gotten through while travelling from one dry place to another.

When you are required to be in it all day, it takes on a different relevance. I don't mind being in the rain all day. It can be immensely enjoyable, and the warm, slightly humid spring rain we are currently experiencing is delightful. It's warm enough to muddle around in dressed in little more than a raincoat and gumboots.

This does have its drawbacks though, especially when guests arrive. It's easy to invite them inside, and be halfway through making them some tea before you realise that you have neglected to put on your pants.

It is a fitting metaphor for my existence (the rain, that is, not public pantlessness). Despite the torrents falling from the sky, I am more isolated from water than I have ever been. For the first time in my life, I am, like hundreds of millions of the world's poor, living without running water. It's very quaint.

All of my drinking and washing-up water has to be carried in containers from the water tank. I miss the luxury of the tap. When I do encounter a tap it's like a strange novelty. I use it with a hitherto unfelt sense of reverence. I find I am treating refrigerators the same way.

I have been to places where clean running water was not available, but there was always the option of bottled water, or beer. Having to carry all my water certainly made me aware of how much I was using. It also made me even more aware of the weight of water.

Water is a heavy, difficult weight to carry. It swills around, throws one off balance, and has a tendency to want to leak and spill. In short, I have no idea how the people who carry water immense distances do it.

I watched women in Mali carrying water in vessels on their heads, and they did so with a grace and poise that was both strong and sensual.

I have no such grace. I lug my water in buckets, or in containers, around the farm, and if I were to put them on my head, I imagine the weight of it would crush me. I would close up like a concertina.

At one point, while labouring with the sloshing, unruly weight of the water containers, I labour also for a few long minutes under the misapprehension that perhaps water in Africa is lighter than the water we have here.

Maybe the African heat keeps the water at a higher average temperature, which spreads the water molecules out, thus making it lighter. How else could you explain the amount of water the African women carry around on their heads?

Surely I couldn't be so radically weaker than they are?

It seems I am.

Even the amount of bathing I do is governed by the water-weight scenario. As I want to remain without too much of a heady musk, I need to wash myself. Standing in the rain working up a soapy lather can only suffice for so long. As a system for remaining clean, it also falls apart when there is no rain.

Initially I constructed a rudimentary ablution block in the field. With some old warratahs (the same steel poles used to hold my fence up) I fashioned a frame for a sink, and a stand for an old mirror that was lying in the shed.

It worked well for a while, and it was a great pleasure to be able to stand in the field and subject myself to a vigorous alfresco flannelling. But then a storm came and broke the mirror. I wasn't too concerned about experiencing seven years' bad luck, because how much worse can it get than living in a field in a tent?

As luck will have it, the previous inhabitants have abandoned a cast-iron bath under the totara tree. I intend to set it on some concrete posts, and then light a fire under it to heat the water when I need to bathe.

The Land Rover serves as a useful tractor, especially when it comes to towing things around, so I fasten a strop around the bath, and haul it across the paddock to the point that I have calculated will be the best position for it.

That position may appear to a casual observer to be simply a random spot in the middle of the paddock, but it was determined by a complicated formula:

Desired position of bath =
distance from things that might burn = tree/tent
÷ distance water is to be carried
where distance = amount of exertion I can be bothered exerting
× depth of bath required for satisfactory bathing experience

With additional factor:

How far is too far to walk nude across field
(= estimated average air temperature × proximity of neighbours)?

In retrospect this formula radically underestimates both the amount of water I can be bothered carrying and the distance I can carry it, which are: not much, and not far.

How full I choose to fill the bath is also governed by whether I utilise water from the nearby trough, which receives its slightly murky supply from a pump that pumps it out of the creek at the bottom of the farm, or whether I choose to cart it from the tank, which is full of rainwater collected from the roof of the farm buildings.

The trough has the benefit of being much closer, but full of potential contagion, whereas the water tank is cleaner, but is some distance further away.

I suppose I could just buy a hose.

The reason I don't is because I want to truly experience the ordeal that is the inability to access running water. I feel there is a certain degree of romanticism attached to the Spartan monasticism of the experience.

I have also deluded myself into thinking the enforced exercise will be good for me.

As the ordeal of water carrying becomes apparent, I begin to reason that if I am to live more sustainably, then there is no need to use so much water. Ergo: there is no need to change the bath water every day. After all, there is only one of me and I'm not that dirty. Surely it can last a few days?

I also theorise that if I fill the tub to the brim, a certain amount of water will be displaced every time I lower myself in. All I will need to do is replace this water each time I get in, and over a period of days the bath will be almost self-cleansing.

If any of this indicates anything, it's the fact that I have a lot of time on my hands for unnecessary thinking while trapped in the tent as the rain continues to fall.

In the television programme I am often seen bathing in my swimsuit. This only really happens when the crew are there, in order to spare both them and the viewers from witnessing me in a more natural state.

At other times I get quite used to strolling from the tent to the bath in the buff. While I don't do it too often when it is light, once dusk comes the field is mine and there is little danger of prying eyes peering in.

Even if the neighbours are looking, I judge the distance to be sufficient that it will be difficult to tell if I am in a bathing suit or not, unless they have binoculars, or a telescope. If they do have either of these, then good luck to them. After all, there's surely no crime in bathing in the middle of one's own field as nature intended, is there?

I have toyed with the idea of erecting a screen of some kind around the bath, but in the end decide it is unnecessary. I prefer the view. Whether the neighbours do, I shall never know.

Over the course of the experiment, the pleasure of a steaming hot fire-bath never wanes. It is always there in the paddock, waiting to soothe aching muscles, and to wash off the dirt and sweat and blood.

Lying in it with all but my face submerged, either just on dusk or late at night, the steam rising off the water and mingling with the smoke, cocooning the bath in a layer of heat that creates a warming thermal layer extending a foot or so above the water, and looking up at the stars or the clouds, is utterly pleasurable.

All is quiet, bar the sounds of cattle and night birds, the occasional swoosh of traffic on the nearby road, the intermittent sound of neighbours laughing (hopefully not at me), and the occasional unidentifiable night sound that is simply not worth wondering about in case it is something scary.

I am certainly looking forward to the bath's warm embrace whilst we are out whitebaiting. The rain has been unrelenting, but Scotty is an interesting character, and the chance to simply sit and exchange tall tales is a welcome break from working for a living. We sit, staring at the creek pockmarked by the incessant drizzle, and wait.

It would have been nice to have a snug whitebait stand, but even they have succumbed to the clutches of the property speculators. In recent years, stands have gone for record prices, bought by speculators and fanatics for exorbitant prices that the locals can only dream of.

Every so often we emerge from the faux shelter of the trees and check the net.

While the glory days of the huge whitebait catches have mostly passed, there have been reports of a Hokitika whitebaiter netting 300 kilograms off a single tide just a week ago. With the delicacy variously priced from $70 to $150 per kilogram, this represents a potential day's earnings of anywhere from $21,000 to $45,000.

I know the chances of us catching that much are slim, but it is always a possibility, although that may lead to a skirmish with Scotty over the rights to the haul. I'm not sure of the etiquette in such a circumstance.

As I pull the net from the water my heart leaps. Its weight indicates that there is something in it. Jubilation is halted when it appears I have caught little more than a stick. However, on closer inspection, there do seem to be a few tiny, pale whitebait in the bottom of the net. Reaching in, we fish the few fish that are in there out, laughing at what the old timers would have thought as we fossick around after single fish trapped in the net's many folds.

In the past, people talked of vast shoals of the tiny fish darkening rivers as they made their way upstream. People would fill kerosene tins with them. Quite why they always seemed to use kerosene tins, I do not know. Whitebait were so plentiful that they were thrown onto gardens as fertiliser, or fed to chickens to such an extent that the hens' eggs were said to taste fishy. Much of this waste was attributed to the fact that for many years there was no real way of storing the fish.

There are some great photos of some of the early pioneers of the whitebait trade, loading tiny, fragile-looking planes full of tins of West Coast whitebait that was to be flown out over the Southern Alps. Those trips must have been mightily perilous, and I suspect those people who eventually consumed the whitebait probably knew nothing about them.

We certainly haven't caught enough for matters of storage or transportation to be an issue, and the most danger I may encounter, other than falling into the creek, is to incinerate myself with a malfunctioning gas cooker.

Shielding the matches from the rain, we light the stove under the shelter of the dripping trees, heat the fry pan, and throw in a little butter. As I have only brought eggs laid that morning by my chickens, all we really end up making is a basic whitebait and egg scramble. I have also forgotten to bring any utensils, so in lieu of a spatula I use a stick to stir the mixture around. Luckily I had caught an ideal one earlier.

The stick isn't much use when it comes to trying to lift the piping hot mass to our mouths, so Scotty and I are forced to share the lid of the butter container to eat it.

It's delicious. The whitebait couldn't have been any fresher unless I had put my head under the water and attempted to suck them into my face.

And still the rain falls.

16

The mice will play

I am having an uncomfortable stand-off with a mouse. Or mice. I'm not sure there is only one mouse, for despite never having seen more than one at a time, I can't be sure there aren't more, as they all look essentially the same — mouse-like.

I have moved out of the tent and into a caravan, a luxury that I have paid for by swapping my Hereford calves, Campbell and Sainsbury, for a loan of it for six months. I hadn't anticipated it would bring with it the night terrors I am now experiencing.

I am sitting at the table in one end of the caravan, and the mice are at the other end on (or worse, in) my bed. I am not sure what to do about it.

I don't know why I am so fearful of rodents. My mother would say, 'Don't be so silly, they are more afraid of you than you are of them.' If this is true, then the mice sure have a funny way of manifesting abject terror.

It's not so much that they frighten me. It's more that I feel uneasy about them. Maybe it's their darting speed, or their sharp, pointed teeth. More likely it's their tail. There is also the reputation they have as vectors of disease. While rodentophiles will argue that they're a clean animal, there's no forgetting that they did bring such events as the Black Death to humanity. Sure, it was actually transmitted by fleas, but the fleas lived on the rodents, ergo, the rodent is a dirty bearer of bad tidings.

I have been trying to trap the mice by placing plastic mousetraps around the caravan. I place them in corners. I leave them in cupboards. I put them under the bed and on the bed where they seem to be poking their heads out. I've put a little camembert that I had kindly been given by a cheese vendor at the Matakana market on the traps. Once again my prey is being lured with better food than I have.

Before I go to bed, I have to tentatively go through all the bed blankets and sheets. There are a lot of them, as the caravan is totally devoid of insulation. It's a nightly routine, lifting the bedding up and shaking it out, hoping the mice aren't clinging to it with their grippy little feet.

I think they might be living in a space beneath the bed, so I seal the gap up with gaffer tape.

We are definitely suffering from a Bible-esque plague of rodents — there is no question about that. In the converted garage, attached to the old farmhouse, and now used as the editing suite, Brendan the editor has been trapping them for weeks. He has disposed of over forty so far, and their numbers show no sign of abating.

Our infestation is nothing like the one that occurred in Australia in 1993, where a perfect storm of conditions led to what has been described as the world's worst mouse plague. Mice numbering in the hundreds of millions overran whole areas. Pigs trapped in their sties were eaten alive. Rivers of mice would flow through doors when they were opened. Even babies were attacked. When the government finally stepped in and a poison campaign was launched, the six-month-long plague was stopped in its tracks in a matter of days.

Instead of building better mousetraps, Australia is now considering using a genetically mutated herpes virus to stymie the mice. Quite how they'll give herpes to mice, I do not know.

Footage of the plague remains some of the most frightening images I have seen on the Internet. Needless to say, each time I see a mouse pop its quivering little nose out from beside my mattress, memories of this come flooding back.

Mice may be the least of our worries though. Something lives in the roof of the old farmhouse too, and Wayne the other editor has to cope with the clomping of whatever it is that lives above him. You can almost set your watch by it: at 5 o'clock every day there is the sound of an almighty THUNK, followed by the thumpthumpthumpthumpthump of footsteps in the ceiling. All we can do is stare at the ceiling and hope that whatever it is doesn't fall through.

Outside the house there are glossy-coated, bright-eyed rats poking their heads out everywhere. They even cavort in the daylight, such is their lack of fear.

Only last week I was standing in the paddock, enjoying the late afternoon sun, when two plump rats chased each other around my feet in a blur of frolicking fur. They bounced through the grass, leapt over each other, and appeared to be enjoying the hot summer sun as much as I was. Sometimes even pestilence can be cute.

Oddly, I was less worried about rodents when I was in the tent, which was no hermetically sealed environment. It was open all round, and should some animal of the night wish to slip in and investigate, there was little to stop it doing so.

The reason the caravan worries me more is for the very reason that it is a sealed environment, which means that if something does get in, it is more difficult to usher it out.

The caravan, green with a white canopy, is comfortable but suitably basic. The owners have installed plush, new red curtains. When they are closed it's like being cocooned inside a cosy womb. The big red fluffy blanket I have only serves to accentuate the feeling. It is comfortingly maternal. Each morning I emerge, feeling reborn.

It is almost totally devoid of wood, with Formica doors on the cupboards, and lots of linoleum.

There are some cupboards mounted along the wall at eye level. These have doors that open downwards. I can never seem to close them properly, and they have a nasty habit of falling open on my head.

The worst offender happens to be the door right above where I sit at the table. It regularly falls open, and its crisp, hard edge scones me right on the noggin. Why I never get around to fixing it, or easier still, simply sit on the other side of the table, I do not know.

I think I get fixed in my ways, and spend an endless succession of long, cold nights nursing a sore head, wondering if this fixatedness isn't a sign of some kind of mental illness. It is a potent and painful reminder of my inability to keep on top of minor repairs, and as such is deeply symbolic of my failure to achieve any of the things I have scrawled on my comprehensive List of Things To Do.

In the roof there is a skylight that opens to allow some much-needed fresh air to circulate. I leave it open pretty much the entire time, even as the nights get cooler, in order to maximise airflow.

I need all the fresh air I can get. Laundry is not something I do a lot of, as it is a little difficult to do. While washing socks and undergarmentry in the bath with me is easy, bigger items tend to be stored in a series of carefully constructed piles on the floor and then neglected (although washing them is on the aforementioned list). I may wear the clothes slightly longer than I should, but I never notice it as I am living inside the aroma.

I also enjoy the health-giving properties of the cool night air, due in part to the half-recalled teachings of Scouting founder Lord Baden-Powell. He always advocated sleeping with a window open at night, as this would prevent one from catching a cold.

He also touted the benefits of a rough towelling after your ablutions. This is a terrific idea, and one I can heartily recommend. Not for me a soft, fabric-conditioned towel. The rough towel, used correctly, provides a delightful exfoliation of the skin, leaving one invigorated and pink. It's jolly revitalising, and after you have stood naked in a small plastic tub of warm water in the middle of a cold paddock, giving yourself a thorough soapy flannelling, the rough towel provides an almost intoxicating finale. At the very least the warming friction keeps hypothermia at bay.

While I expected a certain amount of dampness in the tent, it is the caravan that leaks more. Water drips in from the skylight, and trickles in above the door and around the windows. This leakage is mildly irritating because it makes my slippers damp when I pad the short distance from the table to bed.

Apart from being cold, damp and mouse-infested, the caravan has the added advantage of featuring several things that I am unable to use. These include light bulbs, power points, and even a small fridge. All are useless, as electricity is a joy forbidden to me. They serve as a tangible tribute to a world I have left behind (as if I need reminding).

Whereas others may choose to civilise their caravan, I keep mine in a raw and rugged, frontier-like state. It's very much a man's world. Tools lie strewn near the door, out of the weather — hammers, toolboxes, even on occasion chainsaws and drills. Work boots are kicked off inside so that they won't get wet. Piles of work clothes litter the floor. An industrial-sized container of sunscreen sits on the bench, waiting to be slathered liberally over exposed skin before I venture out.

The manliness of the environment prompts a conversation with a visiting lady friend about the civilising role of women on the frontier, with special reference to their role in the mythology of the old Westerns.

'Women', she says, 'represented civilisation, in opposition to the cowboy, whose untamed manliness echoed the wilderness of the frontier. Women brought with them teacups and linen tablecloths. Men moderated their language and actions around women.'

I have a nagging feeling that she is trying to make a point about something.

There is little evidence of any womanly civilising in my world, despite Jane's best efforts. I like to think of my caravan as an archetypal single man's dwelling, existing as I do on the borders of society, on the frontier, positioned on the furthest outreaches of the civilised world and the unexplored never-never.

I consider myself to be the post-modern personification of the iconic New Zealand Man Alone — all of this in a paddock surrounded by million-dollar lifestyle blocks and a golf course.

The frontier isn't where it used to be, it seems.

I blend into the environment, rather than attempting to civilise it. Whereas other people see a landscape cluttered with drums and timber and half-finished projects, I simply look past the clutter and see the landscape.

Of course, I am not exactly alone. I have the mice. My battle with them is hindered by the weapons I have at hand. The traps I am using don't seem to be very efficient at killing the mice. They just grip them by the nose or the leg.

After I have gone to bed, I will hear the trap go off in the dark, followed by the scuttling feet of the trapped mouse, which means I have to get up, turn on the light (assuming I can find the torch) and deal with a live mouse.

Sometimes I will throw them far out into the paddock, thinking that if they survive the fall, then all power to them. Mostly I have to put them down. This presents challenges. What does one use to kill a mouse?

I use a small curved sickle that I have rescued from the trash some chums were getting rid of. They were attempting to move decades of family clutter from their mother's home, and it was heartbreaking to see what was being thrown away.

It was the old tools that were the worst of it for me. Old tools have a magic to them. They are solid, well balanced, and dependable.

Their family collection included pick-axes, adzes, forks, saws, hammers — it all seemed such a tragic waste, and a sad loss to the country's tool-based history. Sadly I wasn't in a position to take them all with me, so I took only the sickle.

With its rough wooden handle and curved blade, it isn't specifically suited to the task of executing mice. But it seems a better option than the hammer.

The last thing I want to do is to hit a mouse on the head with a hammer.

Actually, that is the second-to-last thing I want. The last thing I want is to have a mouse in my bed.

No. Wait, that's actually the second-to-last thing I want.

The last thing I want is a rat in my bed.

17

Floundering

'How much net's out there?' I ask, trying to sound useful.

'About four hundred metres.'

'And you want me to haul it in by hand?'

'It'll toughen you up.'

The wind-battered, chain-smoking slab of belligerent humanity that is the master, commander and entire crew of the good ship *Donna* stares sceptically at me as I struggle to haul the net in over the bow.

'Can I sing a shanty while I'm hauling?'

'Shit no. Gawd.'

There are few people who fit the phrase 'larger than life' quite as well as Swampy. He wears it as comfortably as the ragged camouflage waders he's clad in. Even the vast Kaipara Harbour seems almost too small to contain his personality.

The *Donna* is no floating pleasure palace; it is four metres of utilitarian aluminium functionality. Some fish guts, the remnants of a previous catch, still swirl around in the murky water in the bottom of the boat.

Rolling yet another cigarette, Swampy says: 'They're a bloody great little fish, the flounder. They can change colour, y'know, depending on what's on the bottom.'

As interesting as this titbit of information is, it doesn't serve to endear me any more to the flounder. I have never really been a fan of them. Someone must love them though, for day after day Swampy is out on the harbour, casting his nets and hauling in the slippery little bottom feeders.

The flounder swim close inshore at high tide to feed, then when they head back out on the tide Swampy is waiting for them, day after day, governed by the wind and the tide, and the mountain of paperwork and regulations and quotas that mark the state of the fishing industry today.

He still has the refrain of the fisherman: there's plenty of fish out there. 'They take a thousand tonne of fish out of the Kaipara every year,' he says.

'That's a lot of fish,' I exclaim.

'It bloody well is,' he replies.

Swampy represents the last gasp of the small players in a fishing industry that used to thrive along the coastlines of New Zealand. Nowadays only a few lonely boats litter harbours and ports where once the great fleets rested.

If there is any industry that has struggled with sustainability, it's the fishing industry. From the beginning of European interaction with this country, the sea around our isolated island nation has been regarded as a source of endless bounty and untold riches that were there to be exploited as briskly as possible.

As I heave the net in over the bow I'm surprised that there aren't as many fish in it as I had imagined there would be. It's hard to see how a living can be eked out in this manner.

The fish that are there are enmeshed in the fine silver thread of the net. I reach for the first of them tentatively, trying to grip it delicately between my thumb and forefinger. It's slippery and firm, and I'm worried about damaging it. I am struck by the knowledge that, at this moment, I am both literally and metaphorically floundering.

I swiftly realise that the only way to deal with the flounder is to firmly clasp them behind the head and wriggle them free from the net. They are well and truly tangled, and it takes a deft wrist to prise them free.

As more of the net is hauled in, the large black fish bin slowly fills with flapping flounder.

Motoring across the harbour to where Swampy has another net set, I'm perched in the bow to keep the weight evenly distributed. The *Donna* surges forward in response to Swampy's smooth push of the throttle, the wind catches my hair, the light shining on the water dazzles my eyes, and sea spray washes back into my face as a stiff breeze tugs at our collars. It's a great day to be out on the water.

It gives me time to soak up the scenery. Low mangroves give way to rolling hills of pasture. Here and there, houses dot the hills between the patches of bush that remain. On one point there are signs of an exclusive new subdivision going in, roads that lead nowhere, and it's all very civilised, a far cry from the conditions encountered by those who found themselves here at the beginning of European exploitation of the sea.

One of the things that Cook and his crew noted on their early voyages here was the sea's bounty, specifically the abundance of sea mammals. Word of their abundance spread throughout the South Pacific faster than the venereal diseases the European crews left with the natives, so that from the early 1790s onwards, people flocked here to make a killing.

Imagine, if you will, that you are standing on a rocky foreshore. You're watching the ship that deposited you there sail off over the distant horizon. You look back to the impenetrable bush that presses down onto the beach, threatening to push you off your tiny foothold. You've heard the tales of the horrors it contains, stories of tattoo-faced man-eaters only too willing to pluck out your eyes and feed them to their children, while their womenfolk eat your liver, the blood running down their faces. You turn again, and in front of you is nothing but the desolate, empty sea.

For now there are only two tasks: the first is simply to stay alive, and the second is to kill and skin every seal in sight. And there does seem to be rather a lot of them. As you set about systematically harvesting them, a nagging fear lurks in the back of your mind . . . what if the ship never comes back?

This was the reality for the first Europeans who came to these far-flung islands at the end of the earth to harvest the bounty of the sea.

During research I conducted for my show on the history of the South Island, I happened upon the story of four men, who in an attempt to escape from the penal colony that was Australia were discovered stowed away on a sealing ship prophetically named the *Adventure*.

As they were well out to sea when found, the captain could do little with them but initiate a 'work for your stay' scheme, until it was discovered that with four extra mouths to feed, there was too little food to last.

It was suggested to the men that they might alight at The Snares, a tiny island group some 200 kilometres south of the South Island. Armed with little more than some rice, an iron pot and a few potatoes, they were rowed ashore to busy themselves clubbing seals until the ship returned to pick them up.

Between them they killed and skinned 1300 seals, which seems like quite a lot, until you consider that they were there for seven years.

At that point, an American whaling ship, the *Enterprise*, happened past, and its crew were shocked to be confronted by three ragged hairy men leaping from the bush.

According to the three survivors, the fourth man had gone mental, and as care facilities for the mentally ill were scarce on the small island group, they had decided that the best way to treat his condition was to throw him off a cliff.

Despite my ineptitude at fishing, I am pretty sure Swampy isn't going to maroon me on the desolate coast of the southern Kaipara, but fishing is a dangerous activity. There is always the danger of being impaled by a fish hook, or drowning.

Despite the proliferation of television ads urging me to do so, I am not wearing a lifejacket. It is a little worrying, especially considering the Kaipara's history as a maritime burial ground. It's notorious for drownings. The entrance to the harbour has several treacherous bars (and my experience as a landlubber suggests there are more than a few perilous bars located in communities around the harbour's edge).

One of the bars at the harbour's mouth is known as The Graveyard, and is the site of somewhere between 43 and 110 shipwrecks. The fact that no one is really sure how many there have been is disquieting. When a place has so many wrecks that people lose count, it often develops a certain mystique. Here though, it is as if tragedy has become so mundane that people no longer pay it any attention.

Should we come a cropper, I am sure that my body will wash up somewhere along the Kaipara's staggeringly long 3200-kilometre shoreline. As someone who doesn't spend a lot of time at sea, when I do I always feel a little nostalgic about our maritime history and the immense dangers and risks associated with it. Sitting just off the beds of mangroves that line so much of the Kaipara, with the gentle waves flip-flipping against the side of the *Donna*, a seagull drifting lazily on the wind, it's a far cry from the dangers faced by the early sealers and whalers.

At its best, sealing and whaling was filthy, squalid and downright perilous work. It required long hours toiling in all weathers, seven days a week, with little to no safety equipment, knee deep in offal, on ships that were infested with rats, with nothing to look forward to but profit or death, mostly of the agonising kind.

In some ways it's remarkably similar to filming *Off the Radar.*

In the late 1700s, Europe ran on the fat of the sea. Whale blubber, sliced off and melted in vast vats, provided the oil for lamps and candles, street lighting, and the lubrication of machinery. Whales produced a very fine oil that could be used to lubricate the mechanisms of the new-fangled pocket watches and clocks. In many ways we could say that the whale was the innocent victim of the tyranny of the timepiece.

Whales were also used in the manufacture of other vitally important things, such as the 'boning' or framework in women's corsetry. Juice extracted from their heads was used in perfume, which in Europe at that stage was an essential, as the concept of taking a bath was considered a bit queer. As for the meat, that was allowed to rot.

Unsurprisingly, the populations of seals and whales declined rapidly.

Such was the shortage that a diarist on an Australian whaling ship visiting our shores wrote that he witnessed one whale being pursued by 70 whaleboats. It wasn't only the whale that was endangered in situations such as this. As the whale was claimed by the crew that first harpooned it, those in the leading boats had to be careful they weren't themselves impaled by harpoons thrown in desperation over their heads by trailing crews. That'd make for a rough day at the office.

It wasn't until 21 December 1964, the same year that *Coronation Street* was first shown on New Zealand television, that the last whale was killed in New Zealand waters. Over the previous years, fewer and fewer whales had been sighted, which was not surprising, because at the same time in the waters of the Antarctic, Japanese, Russian and Norwegian vessels were engaged in wholesale decimation of the remaining stocks, killing tens of thousands.

Ernst Dieffenbach, the first trained scientist to live and work in New Zealand, summed it up best when he wrote: 'The shore whalers, in hunting the animal in the season when it visits the shallow waters of the coast to bring forth the young, and suckle it in security, have felled the tree to obtain the fruit.'

He could have been describing the history of our fishing industry in general.

Nowadays it's not the whale that is harvested, but the money from tourists' wallets as they queue for whale-watching boats to view these mighty leviathans.

The one time I went whale watching it was one of the most underwhelming experiences of my life to date. We did see a whale, but there was none of the breaching, or slapping of tails, or any other frolicking that you generally see on television. Instead we watched a whale sleeping. It slept on the surface of the ocean for a quarter of an hour, then disappeared again for another 45 minutes, before it reappeared for another nap.

I mention sealing and whaling for the simple reason that these industries set a precedent for the harvesting of the sea's resources, where almost every successive venture has featured the exploiting to near-extinction of virtually every species deemed useful. From seals and whales to crayfish, oysters, snapper, toheroa, and any number of other species, supposedly unlimited resources have quickly became very limited ones.

It wasn't always obvious this would be the case. Given our current propensity to consume fish, it is hard to believe that when the Europeans first arrived they chose not to eat it.

Aside from the obvious problems of storing and transporting fish in the days prior to decent roads and refrigeration, in England only those who could not afford meat would eat fish. Eating fish here was considered to be a reminder of the poverty they had escaped, and generally they eschewed it for red meat, which was cheap and plentiful. When they did eat fish, it was in forms they recognised: imported salted herrings and canned fish from the UK. It seems that food miles were not an issue then.

It's ironic that given the high prices of seafood today, a diet high in fish will make you poor.

As stocks of one species become depleted, other species are targeted. At one stage attention turned to a fish that had been so loathed by fishermen that they referred to it as the 'slimehead'. The Japanese were even more disparaging, christening it the 'diarrhoea fish' and refusing to eat it. However, when some bright spark realised that the fish could be made edible by removing a waxy layer from underneath the skin, it transformed itself from an annoying by-catch to the most expensive fish on offer.

Rebranded by marketing folk as the 'orange roughy', it was hauled from great depths in vast quantities, until it was realised that the fish was a slow breeder, taking up to 20 years to reach maturity, and that breeding stocks had been decimated. It is now considered by some to be in the worst state of any of the fish stocks we harvest.

I am glad Swampy and I are not on the hunt for diarrhoea fish.

I watch as he guts the flounder in the blink of an eye. I try to follow, but it's all a bit too fast. A quick slice with his knife, a deft flick of his wrist to extricate the guts, and it's done.

'Don't just stand there gawking, get rid of those snotties,' he says.

The snotties are the jellyfish that are caught in the net. As their colloquial name suggests, they look as if someone with a rather bad cold has sneezed into the net. There are dozens of them. Cold and mucusy, they slosh around underfoot until I grab a bucket and bail them out over the side.

Turning to roll another cigarette, Swampy looks at me and says, 'Well, you're not a complete disaster.'

High praise indeed.

18

The goats are wild

It's not often that you stop what you're doing and think to yourself, 'You know, I hadn't really anticipated when I got up this morning that I'd end up with my hand stuck in a goat.'

I am not a fan of hunting goats. Don't get me wrong — it's not that I don't think they should be hunted. It's just that I have always found them incredibly frustrating to hunt.

Goats love hills. I do not like hills. Goats have no problem scaling them at great speed and with remarkable nimbleness. I can barely drag myself up them, sweating and huffing and crawling on all fours.

Matt the butcher, who I last saw when we went pig hunting, has offered to take me to one of his mate's farms to hunt goats.

John's place turns out to be one of the nicest farms I have ever seen, a large beef-raising block that wraps itself around a tributary of the northern Kaipara.

The block has a small mob of feral goats that were present when John moved there, but he's since brought some larger goats from up north and introduced them to the herd in the hope that romance will blossom between the mobs. The resulting kids will be bigger and meatier. He continues to allow the goats to range freely, shooting a few whenever they need dog tucker.

Along with the goats, there are roaming mobs of feral pigs, and John ensures that a healthy breeding population remains by having safe havens where hunting is banned, so that the sows can raise their litters in peace. Outside these areas, he and his buddies can hunt them whenever they need more pork.

Sustainability theorists will say that the game is fed on solar energy because the farmer adds no other ingredients; the sun and rain provide the foliage for the goats and pigs, the dogs eat the goats, and the pigs feed the families all over the district.

And now some of it is going to feed me.

I have only ever killed one goat before, and that was a year earlier. To be honest, I didn't personally kill it; a Dogon medicine man in a village in Mali did the actual slaughtering. All I had to do was tell him what purpose I wanted the goat sacrificed for, and pay for his shamanic services and the goat (although I suspect the New Zealand taxpayer actually paid this, as the cash came out of the *Intrepid Journeys* budget).

As we stood in his small courtyard, surrounded by various shrines and totems, he held the goat down and asked me what it was I wanted from the sacrifice.

I said I wished not to be killed by a hippopotamus and for good ratings for the programme.

These may seem trivial to you, but I was about to head up the Niger River for several days, and the hippopotamus is the most dangerous animal in Africa (if you don't count the mosquito, which kills more people than anything else).

The priest murmured an incantation, then slit the goat's throat.

There sure is a lot of blood in a goat.

Afterwards the medicine man got to keep the carcass of the goat, as its flesh is considered sacred and could only be consumed by him and his medicine-men chums.

Those of you without any belief in the protection afforded by this ritual may scoff and say that all I did was pay for their dinner, but I wasn't killed by a hippopotamus, and the programme rated very well, so maybe the goat didn't die in vain.

Who can tell? You can't be too careful, I say.

There is very little in the way of ritual while we are hunting goats at John's farm, other than my application of sunscreen. Sunscreen always worries me. I have no idea what it contains, but I slather it on by the litre, as my fear of sun-inflicted cancer is greater than the possibility that I may contract something from all the chemicals in the sunscreen. It simply seems the lesser of two evils, especially as the sun only needs to rise above the horizon and I become a lighter shade of puce.

It is certainly a day that requires sunscreen — a glorious midsummer day that envelops us in a hot, dry heat, the kind of heat that screams GO INSIDE AND LIE DOWN AT ONCE! It is most certainly not the kind of heat in which I wish to scale hills.

In order to minimise the hill climbing, I am equipped with something that will hopefully negate the need to be too close to the goats, namely a large-calibre rifle containing steel-jacketed shells that pack a mighty wallop over a fair old distance.

John thinks the goats will be near the back of the farm, grazing in the shade of the mangroves. Sure enough, from the brow of the hill we can see a small herd meandering through the trees at the water's edge, grazing as they go.

John, Matt, Matt's two boys, Philippa the camerawoman, Frank the soundman, Jane the director, and I all sneak down the side of the hill. We are hardly the most surreptitious bunch, and by the time we get to where the goats have been, the goats have gone.

John selflessly climbs the steep hill again to see if they are over the other side, signals that they are, and then the ordeal I suspected would happen does happen. What follows is an infernal hour of climbing up and down precipitous hills in an attempt to head off the goats.

Leaving the crew, I clamber up the slope as John gestures from a wearingly distant ridge that they are just over the next hill. This comes as no surprise to me, because in my experience goats are always just over the next hill.

Popping my head up above the ridge, I catch a fleeting glimpse of the gently swishing tail of a goat vanishing around the next bend. Up and down I drag myself, but the goats prove elusive. It is made worse by the fact that they aren't even climbing the hills, but sauntering along the bottom.

On the positive side, the scenery is magnificent. From the ridgelines I can see the Kaipara stretching away south like a great silver smear. The river that borders the farm is like a mirror. Birds sing their songs of summer. Cicadas chirp. Cattle stare at me as I pass them. Blowflies drone lazily around the cattle's dung and my feet, which seem to have an inability to avoid said dung. The only thing that tarnishes the tranquillity is the thumping of my racing heart and the ragged, gasping breathing of my lungs.

I suspect one of the reasons this country revered Sir Edmund Hillary so much was that none of us can be bothered climbing hills, and we were happy to bask in his reflected mountaineering glory.

Finally I am in position. At least I hope I am. I honestly believe that if I have to climb another hill, I will pass out. This will be bothersome, as I will probably roll back down the hill in my unconscious state, and then I will have to climb all the way back up again.

Sitting in a hollow near the brow of the hill, I wait, basking in the sun, as the goats clamber up the slope towards me.

After a short while, the faint sound of goat snickering drifts up the hill, and within moments the first head appears over the rise, about 70 metres from where I am lying. Within seconds there is another, then another. They are no doubt enjoying the summer day as much as I am, unaware they are in mortal peril.

There is always a delicate balance in situations like this. You have to decide how long to wait, and how close you can let them come towards you, in order to make the shot easier and increase the chance of bagging more than one, before they catch wind of you and flee.

There is also the question of which of them I want to shoot. I abhor this part of the process.

I'd hate for you to think that this is an entirely enjoyable experience. After all, the goat, like the other animals I am required to hunt, is a beautiful creature. They are so cute and funny looking with their big floppy ears and their swooshy little tails and the quizzical looks in their big yellow eyes.

I certainly don't want to bag an old buck, as they are tough to eat, and their manliness pervades their flesh, rendering it pungent and foul tasting. There are several large, tender-looking kids, and a few small females in the group. Sadly for them, I need to eat, and they have the misfortune of being the best eating.

Spread-eagled under the sun, the dry grass aromatic but prickly beneath me, I scan the group through the telescopic sight of the rifle, trying to figure out which goats I want. Normally I would be satisfied with just one, but I have plans to secure a couple for a special project.

In the end I simply decide to go for a good-sized nanny, and also for a large white kid with a ginger splodge on its side, because it looks as if it will make the best rug.

Taking careful aim at the female, I squeeze the trigger, and watch as she bolts forward several paces, stops, looks about quizzically with her big yellow eyes, and falls over.

As the sound of the shot reverberates across the hill and the mob erupts, I shoot the little ginger goat as it turns to run. The rest of the goats disappear over the hill in a cloud of dust and flicking tails to live another day.

All that remains is to dress the goats. Or undress them as the case may be. In the cool of a grove of trees, we throw a rope over a branch and haul them up.

Skinning them is a rather physical task. At one stage, as I insert my fist between the goat and its skin in order to punch it free, my hand gets jammed. Struggling to free it, I nearly fall over and end up hugging the swinging carcass of the goat. Matt finds this hilarious. He's already mocked me for shooting the smaller goat in the butt. I didn't. I shot it through the rump as it turned to run away. The rump. Not the butt.

With the carcasses skinned and gutted it's back to Matt's butchery to render them into their constituent parts. I want boned roasts, shanks, chops, and a large pile of the rest of the meat for my special project.

With the goats broken down, I leave with them in boxes. They are certainly much easier to handle this way. And the total bill for the entire procedure? Two goats shot and butchered for the princely sum of a lunch at the farm for Matt and his family.

A few days later, armed with a pile of goat meat and a large pork haunch in a bucket, I'm knocking on the door of Salumeria Fontana in Hoteo North, southwest of Wellsford, where master butcher Greg Scopas lovingly creates gourmet sausages.

Greg has volunteered to help me turn my goat into something a little more exotic — salami.

'Have you ever made goat salami before?' I ask him.

'No,' says Greg, 'and depending on how this works out, I might not make it again.'

I can understand his sentiments. There is an unmistakable, almost overpowering aroma of goat in the air. It has a musky feralness to it, although after a day's exertions the same could be said of me.

When I moved to Henderson a few years ago, a chum and I were comparing butchers, and he'd remarked that he'd heard of a fantastic sausage-maker whose premises were located somewhere near my house. I never managed to track down this Shangri-la of sausages, and it had been a source of much annoyance.

As I get chatting to Greg, I discover that he is that elusive sausage-maker.

How serendipitous.

It is little wonder I couldn't find him as he had relocated to his olive orchard in Hoteo North, where he now runs an olive press and his sausage-making empire. The sausages are made by hand and are robust and meaty, based on the French, Italian, and Spanish style of sausage.

Greg's is yet another of the many businesses I have encountered whilst filming *Homegrown* and *Off the Radar* that began as a passion and progressed to successful enterprises. He'd begun making sausages as a hobby, and had heard that someone in Henderson was selling a commercial sausage-making machine. It turned out to be located at a disused sausage factory, so he bought the machine and the factory.

'Do you ever eat a sausage-sizzle sausage?' I ask him.

'Sometimes I give them a try,' he says with a laugh.

'Ever thought of doing a gourmet cheese sizzler?'

'No.'

Over the course of the next hour we mince the goat and pig together, which unbelievably serves only to increase the heady aroma of the goat meat. This mince is then combined with garlic from Invercargill, French chilli powder, Sicilian sea salt, South Australian wine, and a few other spices, and crammed into a metre-long piece of pig intestine. As great as the salamis look and smell, all I can see is my carefully reduced food miles going out the window.

Hanging the multitude of salamis that we have fashioned in the curing room, Greg explains the ethos behind his gourmet sausage and olive oil business. 'It's pretty simple, Radar. If you want a quality product, you use quality ingredients.'

'Exactly,' I say, 'and what could be better than free-range wild goat?'

'Well . . . pretty much anything, I imagine,' he replies.

Good one, Greg.

19

Birds of a feather

If I were to draw a very long bow, and aim the arrow of symbolism at a pressing social issue, then I could say that the aerial fracas I witness in the skies above my landholding presents an apt metaphor of a wider societal issue, which is the increasingly common clash between the rural and urban populations. The face of the countryside is changing.

From the plateau on which I reside, I can sit in my thinking chair and watch the ongoing skirmishes between hawks and the resident magpies. The magpies, brash and loud, rise and swoop upon any unfortunate hawk that flies into their sphere of influence. The hawks look resigned to it and attempt to fly up and way, but the persistent magpies swirl and attack.

It can go on for several minutes, until eventually the hawk, either wearied by the annoyance, or convinced there are slim pickings to be had there anyway, moves off. The magpies return to the tree to do whatever it is they do when they are not attacking hawks or diving at people's heads.

Magpies are a ferocious bird. When they attack you, you're forced into all kinds of humiliating defensive actions. There is no dignity in being attacked by a magpie. They will make you look foolish to any onlookers, who will laugh at you until they too are attacked by the birds.

With the flood of the urbanite into the rural hinterlands, there have been many similar confrontations as people with unrealistic expectations, or flawed views of the realities of the countryside, come into contact with those already there.

Not long ago, the sound of calves and cows who had just been separated from each other at weaning would not have prompted a call to the council noise control officer.

No one would have considered taking a farmer to court because he milked his cows earlier than the complainant wanted to get up. They would have been laughed out the door. Not any more.

In Kaukapakapa, the local volunteer fire brigade has received complaints by one new resident that their siren, which helps alert the members to the existence of an emergency, is too loud, and please, would they mind not using it at night. One can only hope that person is never in need of the assistance of the volunteers.

This clash is not something that we have resolved. Like the magpie, there is loud and vigorous dissent on both sides.

But it's not all bad.

One quiet Saturday, I pack some produce and head to the Matakana markets to play at being a stallholder.

I've harvested a bunch of artichokes and some lettuce, and I have my bonsai lawns with me, although I have upgraded them. Gone are the ice cream containers. The boxes are now crafted from recycled matai, which I obtained from my neighbour, Demolition Pete, who salvaged it from a building he was demolishing.

Sales of the artichokes are brisk. People who have never tried an artichoke decide they will give them a go. A passing woman buys two artichokes for $7. I allow people to dictate the price they want to pay for them. That way it is like a donation.

How rabidly anti-capitalist of me.

I quite like playing storekeeper. I like the contact with people. I like hustling for sales. There is a simple pleasure in the markets.

I had never been to Matakana before the farming project. It was once a sleepy little town, but now it is the hub of an exclusive enclave of wineries, olive orchards, multimillion-dollar lifestyle blocks, and the most famous of the farmers' markets in the area.

The coffee is superb. A band is playing. It's more than a trip to the greengrocers. It's an event.

People are interested in what we are filming. It's one of the great gateways to conversation sometimes, having a camera crew. I get to chat to stallholders and customers. Over the course of a few more personal visits during the following months, the relationships are maintained, stories are swapped. There is a palpable sense of a village market, despite the fact that people flock here from miles around.

The irony of this festive market is that many people drive the long hour north from Auckland to buy 'local' produce, then return with it to the city. The fact is not lost on the market people.

But does it really matter? It's easy to mock, but maybe we shouldn't. It's a great day out, regardless.

A big family outing for us when we were children was to pile into the car after the cows were milked in the morning and simply go for a drive. On this drive we would look out the window of the car, and watch the landscape roll past. Much of the time I don't even think there was a destination as such.

It was something for which I shall for ever be grateful.

It would be something that many people would find boring, but I loved it. At least I think I did, and I don't want to ask in case I discover that as a child I didn't love it and made the entire experience miserable for everyone. I wish to continue with thoughts of my idealised childhood.

Going for a drive is something I still do to pass the time, even though we now live in an age where the thought of simply driving somewhere for no actual reason is frowned upon as a waste of precious oil. It has also become a lot more expensive to do, but it is a great way to see the countryside.

With the advent of fitted televisual screens in vehicles, passengers now watch DVDs or play games. Their focus is on the internal.

Just in case Te Radar forgets where he is, a handy sign marks his boundary.

Pansy the goat rightly wonders why Te Radar is bathing in her drinking water.

There's little that makes one appreciate electricity like having to light a fire to boil water every time one feels like a cup of tea.

Several concrete posts, some bricks and a cast-iron bath make for a wonderful outdoor hot tub.

Te Radar cooks his lunch in the kiln while Bevan the cameraman and Frank the soundman record the process.

A fire rages in the heart of the kiln, in preparation for cooking more tasty morsels of animals that once ranged in the paddock beyond.

Hay trial garden; little grows but grass, of which there is clearly no shortage.

Sawdust trial garden; even hope fails to grow here.

Soil trial garden; marginal success, which proves little but that potatoes will grow virtually anywhere.

The land provideth a turkey . . .

. . . and the turkey provideth a meal.

Christmas the turkey performs another futile mating dance in front of a disinterested Te Radar, who is engaged in idle contemplation. More idle, perhaps, than contemplating.

Bevan the cameraman catches a moment of evening intimacy in the caravan. Note the squalor.

The kitchen of 'The Chapel'. No women dwell here.

Fruits of thy labours: goat rug, peaches, goat salami, lemons, and apples.

Building code violations are compounded by fashion crimes.

A man in a meadow. The enclave is complete.

The external, the passing landscape, is lost to them except through quick glances.

Having travelled on some major highways overseas, I can well understand why this is. There is no view to speak of.

A good way to monitor the changing nature of the countryside is to watch the speed at which houses and subdivisions encroach upon productive farmland. There are those who decry this, but if people wish to live in the countryside, then who are we to stop them? Who are we to say no to the tiny blocks, the mini vineyards, and olive groves? Or have we reached the stage where, like the magpie, we need to drive them off to another hunting ground?

Another woman approaches my stall and pays $10 for four artichokes. She asks me how they are prepared. This is a question for which I am not prepared. I tell her I have no idea, and we both agree that if we find out over the course of the morning we'll let the other know.

Ambling through the markets is Dougal from Dublin, who's doing a PhD in farmers' markets around the world. It sounds like a very taxing job — all that travelling and mooching about, staring at produce and meeting like-minded folk.

'What is it about them that draws people back to them?' I ask.

'People are becoming more aware of the importance of good-quality food,' he says. 'They want to form a connection with the person and the product.'

'Right,' I say.

'I like your lawn,' he says. 'I have never seen anything like it before.'

I'll take that as a compliment from a man who knows his farmers' market produce.

Joe from Rainbow Valley Farm also likes them. 'People could graze their hamsters on them,' he says.

I wish I had thought of that.

Joe, like most of the other people at the markets, seems to be in a particularly good mood. 'Look around,' he says. 'People are smiling when they are buying things. You don't get that in a supermarket. Here the transaction is from person to person, from heart to heart.'

While it may sound a little schmaltzy, it is actually true.

There is a delight in the markets. People are there to shop, much as they are in any store, but the atmosphere is different — part village market, part tourist trap. Smiles are exchanged as often as money.

There's also an honesty about the place as people sell the wares they have made. They will be back next week and the week after, so if there is anything wrong with their product, they can't hide.

From fruit and vegetables that really do still have the dirt on them, to preserves, pickles, wine, cheese, sausages, bacon and coffee, there is a little something for everyone.

Despite the bonsai lawns looking good, they don't seem to be selling. Maybe they are ahead of their time. I'm sure that if I were to take them to the markets in the city, I would do a roaring trade. It's all about finding the right consumer for your goods.

In all honesty I don't really mind, because they serve another purpose: they provide a great opening gambit for conversation, as person after person stops to stare at them, and then ask what they are for. I while away the morning, talking and laughing with strangers.

Trish, who I have not seen since my visit to Rainbow Valley Farm, arrives and we chat as people peruse our wares.

She admires my artichokes.

'I've sold quite a few of them,' I say, 'and I gave one to a girl with a pretty smile.'

Trish holds one up. 'These ornamental ones are so beautiful,' she says.

'Sorry . . . ornamental?'

'Yes, you can tell by the shape of the leaves — they are more pointed than the edible ones.'

'Oh.'

I think I have been misrepresenting my product. I fear that the woman who has always wanted to try artichokes may not be so keen on them any more.

As I'm about to leave, Joe returns. Putting his arm around Trish he says, 'You know, Radar, if you are really going to be sustainable and you want to live off the land, what you need to survive is a good woman.'

She beams. He grins.

I laugh. 'Which stall sells those?'

It is to be the last time I see Joe. A brain tumour claims his life a few months later.

20

Boys will be boys

For all my concerns about spiders, the most dangerous animal in the woodpile is probably me, especially when I am wielding an axe, for it is fair to say that I am no axeman extraordinaire.

While I will survive intact several months of firewood chopping on the farm, late last year I found myself at the Canterbury A&P show, convinced that within moments there was going to be less of me than there had been.

Shooting the *Homegrown* 'Timber' episode, I was exploring the world of competitive axemen. I have always been impressed by any competitive endeavour that combines a job with a sport.

In most sports there is a danger of injury, but few sports have a piece of equipment that is honed to a razor-sharp edge, and which is then aimed at a large piece of wood only centimetres from where your feet are precariously positioned.

While I knew that in participating I would have to stand on the log, and swing the axe at a point between my feet to cut out a V, I had anticipated that they would provide me with some kind of safety footwear, but they laughed at me and pointed to the light sandshoes on their feet. 'That's them there,' they said.

They were ordinary sneakers.

Ludicrous.

They wouldn't have impeded the axe in any way whatsoever. The axemen may as well have protected their feet with a coating of butter.

'We used to do it in socks — you get better grip that way,' one axeman said.

'What about your toes?'

'Well, you don't aim for them.'

How this is still legal I do not know, but that is the world of the timber industry.

Of all the tasks I attempt on the farm, it is always the ones involving timber that have the most chance of something going haywire. This is because these are the tasks that involve any number of dangerous tools — from circular saws to chainsaws, from hammers to pneumatically powered nail guns, not to mention the sheer weight of the wood when it comes plummeting down towards your upturned face.

Thankfully, most of the work I have to do with timber involves working alongside Demolition Pete. Pete is another chap who just happens to live up the road, and while he currently specialises in demolition, he has formerly been a house mover and a tree feller fella. I first met him when I needed some timber for my gardens.

If anyone has timber, it is Pete. His shed is stacked to the gunwales with all manner of recycled timber. Contracted to pull down houses and buildings, Pete strips out what is valuable and dumps the rest.

As a sideline he uses a lot of the timber to build small transportable buildings. According to the council regulations, if a building isn't fixed to the ground, is under a certain size, code compliant, not slept in, doesn't have potable water, and isn't cooked in, then it doesn't require a permit.

Built largely from recycled timber, Pete's buildings look like little weatherboard houses, complete with windows and French doors. They are grand-looking buildings, and I want one, which I envisage selling at the end to recoup the costs.

So Pete and his offsider Brett arrive with the prefabricated kitset components on his truck.

'Where do ya want it, Mr Radar?' he asks.

'Ah, about there?' I suggest, pointing in a non-specific way at a generalised area of grass.

'Well, that'll give you a nice outlook,' he says.

It most certainly will, as it looks out over the field and down into the valley. People pay a lot of money for a view like that. Of course, the money they pay for the view also generally secures them a house with power and running water.

With the location agreed upon, it is time for a good old-fashioned house raising. Within a short time the floor is on, the wall framing is in place, the roof framing is lifted on, and the beer is being drunk.

Over the next few days, the building progresses smoothly, and within a short period we are nailing up weatherboards. I manage to get mine a little skewwhiff.

'Don't worry about that, no one'll ever know,' says Pete, echoing what I suspect is an often-used builder's refrain.

He even provides a lead-light feature window, and Brett lends me an old deck that he has lying around. These are the crowning glories.

As the building reminds me of the tiny wooden churches that dot our countryside, I call it The Chapel.

It only lacks one thing (if you don't count power, or running water) — a roof over the deck, and while I am at it, one in front of the caravan. I need to break up my indoor-outdoor flow, as the present tendency is for the outdoors to flow a little too liberally into my indoors. I need areas that are neither indoors nor outdoors, but convenient middle grounds.

I had always planned to build verandas, but with a tight schedule, things have been prioritised differently and I have not thrown them up.

But now at last, as summer draws to an end, the time has come to create something that will protect me from the imminent threat of rain that autumn holds. Being hot is one thing, but being wet is a lot more miserable.

Pete suggests we use manuka poles from his mate's place, which I consider a marvellous plan as they are free, and it's legal to harvest a certain amount.

Collecting the manuka is one of those jobs that you think should take 10 minutes, but really needs an hour, and ends up taking far longer than you could ever have imagined in a worst-case scenario.

The timber is located on the far side of a valley, and can only be obtained by descending a long, winding, clay track with a sharp bend at the bottom, just before a skinny bridge over a creek, which leads to a steep slope on the other side that rises to a small dome-shaped plateau, on which you have to attempt to turn around before collecting the timber.

Had the job been done a few days earlier, we wouldn't have suffered the effects of the first of the rain on the dry ground. But with a series of squalls running through, a light smattering of rain has reduced the few inches of topsoil to a soft mire that sits atop a layer of clay, which has received just enough moisture to render it a slick slime.

What could possibly go wrong?

As Pete's trailer is too big and we don't have mine, we borrow a small trailer from his mate. One tyre is flat, and there is no pump. Still, it's only a few poles. How heavy can they be?

What I had assumed would be the light poles I want actually turn out to be pillars, some about 30 centimetres thick and a couple of metres long.

As Pete fells them, I think to myself that they appear to be slightly larger than I had imagined, but as he has gone to all the effort of cutting them down, I don't wish to point this out.

There's still a noble grace in the art of felling a tree. There's almost a denial of the inevitable at the beginning, followed by a slow topple, then an increasing downward surge that culminates in the tree finally crashing down.

It is little surprise that the forestry industry's ACC premiums are so high. Trees can fall a long way, and there's little standing in the way of them slapping you to the ground if you yourself are standing in their way.

The chainsaw is still one of the most vicious and raw of all of the tools. That sharp chain designed to cut timber wouldn't flinch if it were to cut through flesh. The thought of it kicking back and flicking its teeth through head or shoulder, gouging out a swath of flesh and bone as it sinks in, does tend to make one a little more cautious. At least initially.

There's any amount of safety gear to wear now, including chaps made out of a fabric that immediately entangles and stops the chain if it is about to cut into your leg. Not for the first time I wish I was wearing chaps.

In the Kauri Museum at Matakohe, which I consider one of the country's great museums, there is a grand photo. Situated as a backdrop to their chainsaw collection, it depicts a bushman on a log, holding a very large chainsaw and clad in the safety gear of the time, which seems to consist of nothing more than a pair of short shorts and boots.

There seems to be an awful lot of muscled limbs and torso exposed to the potential damage that could be inflicted by what is a very large, quite heavy, primitive-looking chainsaw. It has clearly been taken in the days when men were men and workplace injuries were considered an acceptable occupational hazard.

Collectors are wonderful people. It is my great good fortune that my father is an avid collector of vintage tractors and military vehicles, which he spends years painstakingly restoring. Through him I have developed a belief that one of the greatest sins is the melting down for scrap of so much of our heritage.

There was a time when every primary school and playcentre,

kindergarten and park, especially in the country, had an old tractor plonked in its grounds. Sadly, as they deteriorated with the weather and became unsafe, with rusty edges and sharp corners, they were often sold as scrap. These machines that enabled the taming of the bush and the tilling of the fields, that hauled and dug and lifted, were all sent to be consumed by the fires of the foundry. History melted.

I always experience a pang of horror when I hear of priceless machinery having been sent off to the scrapyard simply because no one knew what it was.

I have had the pleasure of experiencing these machines through associations with vintage machinery clubs, and the men and women in them, for whom antiques are not things to be kept behind glass, but living, breathing (usually clouds of smoke) entities.

The destruction of our artefacts even extends to our architectural history. Whereas Europe has castles and cathedrals, our iconic buildings are the abandoned dairy factories, or the reddish-brown woolsheds, symbols of an economic prosperity long past. Now decaying slowly, or being used for some other purpose, they litter the landscape of the countryside.

Jim Downes, the chap who collected the chainsaws now displayed in the Kauri Museum, had been a chainsaw vendor, and as people had brought their saws in to be traded or disposed of, he thought it might be interesting to hang on to a few. Those few formed the basis of a now extensive and unique collection. Were it not for his foresight, these machines would no doubt have been lost to history too, and that would have been a great shame.

They line the length of one wall of the museum, sitting in silent homage to the men who used them. Where some countries display the swords of royalty and warriors, we have the shining blades of the crosscut saws, the pitsaws and the chainsaws. They are our Excaliburs, grimly doing battle with the endless timber that sprouted from the earth of the bushscape that was New Zealand.

Of course, it is a shame they cut down quite so much of it, but anyone who has actually wielded a chainsaw would be loath to blame our forebears, because as it is well known, once you start chopping into things, it becomes very hard to stop.

This propensity for mass destruction by chainsaw is nowhere better seen than back amongst the manuka, where a large pile of logs now lies stacked near the trailer. We have finished felling, and are surprisingly unscathed thus far. I suspect there is not a tree yet that has stymied Pete from felling it. He has a forest of tree-felling stories.

As the sun starts to sink, the dew creeps back in again, causing the grass to become just a bit too slick to provide any traction for Pete's jeep. The vehicle has a tendency to want to slide sideways down the steep hill, which I take to be a slightly undesirable tendency, as the thought of being involved in a rollover accident in the back of nowhere isn't overly appealing.

It would be inconvenient and embarrassing. There would no doubt also be some pain, especially as I am in the back with a couple of chainsaws and various apparatus that would be thrown round inside the cab with me.

The trailer is looking much sorrier under the weight of the poles. The tyre is completely flat, and is resting on the wheel rim.

Ten minutes pass, and the jeep fails to pull the trailer up the slope to the track, and so is removed and replaced by the quad bike of Pete's mate, who has turned up to help. The wheels of the bike are now spinning in as fast a forward direction as they can go, but neither the mud they are throwing up, nor the smoke being emitted in great grey clouds by the screaming motor, is thick enough to obscure the fact that the bike and trailer are moving inexorably backwards down the slope.

The winch on the jeep is not something that can be hauled out with any speed, and the distance it has to go provides quite a lot of time to stand around and watch as it is dragged out and down to where the bike and trailer have come to rest at the bottom of the hill.

It's now dark.

Ten minutes after the cable is attached to the bike, it's hauled back to the top, and the decision is made to leave the timber on the trailer at the bottom of the hill and come back the next day.

The next day eventually becomes a few days later, but undaunted, Pete and I and Brett return to reassemble the jeep and trailer and try again.

With the trailer being so short and the wood being so long, most of it not only protrudes over the back, but is hauled forward as far as possible, so that we are barely able to turn through the tight corners of the track without hitting the back of the jeep with it.

As the jeep climbs the hill, some of the wood falls off, dragging other logs with it, which is unsurprising as we forgot to bring rope and are making do with a rotten length of strop.

I clamber down to where the logs have rolled and drag them back up. Or try to. They are so heavy it requires two of us to struggle up the hill with them.

Once they're finally back on the trailer, we tie the logs on with the severely damaged strop, and by sitting on the front of the jeep, our extra weight manages to hold the wheels down to gain a little extra traction to round the last bend.

Finally we reach the top of the hill, and with the slope behind us we trundle around to the gate, congratulating ourselves on a job well done. It's then that Brett says, 'Hey, where's the trailer?'

The discerning reader will determine from this statement that the aforementioned trailer isn't where it should be, which is attached to the towbar.

It has come adrift from the jeep, merrily returning to its favourite haunt, the bottom of the hill.

Still, on the positive side, all the timber still appears to be on it.

If collecting the poles was a farce, actually fashioning something out of them is a saga, as I am to attempt the construction alone. In my haste to choose the handsomest manuka from the grove, I have unwittingly caused myself a problem. The poles are clearly too long, and I have no chainsaw to shorten them. The ones I have selected are so heavy that I can barely drag them to the hole into which they need to be placed.

I have dug the hole as deep as I think it needs to be (and as you are now familiar with my digging technique, you will know what my estimation methods are like) and in this case I have estimated quite deep as the pole is quite tall.

I manage to get the pole in without it crushing me, step back, and think, 'Yep. That hole should have been deeper.'

I push the pole straight, as if by straightening it, it will somehow become shorter. Hope springs eternal.

Wrapping my thighs and arms around the rough bark of the pole, I squat a little and try lifting it out. The strain suggests that there is a very real possibility that I could give myself a hernia at any second, or at the very least blow some kind of fufu valve, and my lifting is to no avail. The weight of the pole has caused it to sink into the sticky clay and it is stuck fast.

How tedious.

If the manuka was not quite so hard, I may have considered sawing it off with the handsaw, but it is so dense and the saw so blunt that I would be better off trying to shorten it by attempting to lick the top off it.

Just as I am about to concede defeat, the cavalry arrive in the form of Pete and his buddy Vic. They just happen to have been passing and have decided to call in to see how the poles are progressing. I tell them they are very fortunate, as not only can they now take a look at progress, but they can help progress it themselves.

Thus we manage to haul the pole out and I dig the hole to the depth I have calculated it requires. We heave it back in.

Pete and Vic stare at me as I stare at the pole that is still clearly too high by a lot more than the amount I had imagined, so we haul it out again, and I recommence the digging.

When the hole is well over a metre deep in the hard clay of the hill, we drop the pole back in for what we hope is the final time. Perfect. Or at least as near to perfect as is necessary.

There is nothing for it but to celebrate with a beer. The other poles can wait until Pete drops off the chainsaw so that they can be judiciously trimmed to a more suitable length.

Over the next few days I sink the now sensibly shortened poles, erect some cross members, and spend a lot of time standing some distance back from the structure, wondering if I have judged the angle correctly, or whether the roof will have little or no slope to it, meaning the water will pool in it rather than running off.

I've appropriated from TVNZ a large billboard tarpaulin that has been used to advertise *The Sopranos*, and it will provide the roofs for the verandas. After the saga of the pole fitting, it proves to be a straightforward job, and within a short period I have two wonderful roofs.

Standing back to marvel at my mastery of my domain, I realise that I may have put the poles in a little too close to the caravan and it may now be wedged in.

Bother.

21

White butterflies

It was at the point that I discovered I had perfected the technique for successfully urinating into a watering can, without contaminating the handle or surrounding area, that I realised that taking matters of sustainability into your own hands can be more complicated than I had anticipated.

I had certainly not imagined, when lunching all those months ago with the producers, that it would come to this: a man, a plan, and a watering can.

The reason for this degradation is simple: my proverbial field of dreams has become a paddock of nightmares. My Garden of Eden is barely even a garden. Nothing is growing at anywhere near the rate or volume I'd expected. In fact, nothing is growing. I face the ignominious prospect of gardening failure and starvation.

I had thought that by now I would be growing so much that I would have been trading farm-fresh produce for all manner of things, such as alcohol.

Having been conducting my field trial, I have come to the realisation that gardening is more difficult than I ever imagined. It has always appeared so easy on the television: people plant things, some kind of editorial device is employed to indicate time passing, and their produce is suddenly bountiful. I am beginning to feel as if I have been deceived.

In the hay garden, despite my belief that the heat caused by the decomposition of the bales would give the plants a little boost, all that has happened is that the seedlings appear to have gone yellow. The strawberries have diligently clung on, showing some promising signs, but they are virtually the sole surviving thing growing in the hay that isn't grass. This grass must have germinated from seeds in the hay, and it seems to be flourishing. I'm not sure how much nutrition it is going to provide.

In the sawdust garden it appears as if some of the seedlings have gone into stasis, as after several weeks they haven't grown at all. Some of them look (if this is indeed possible) to be smaller than when they were planted. The only things that appear to be doing well are the chillies, and some plant I can't identify, until someone suggests that it might be an aubergine.

The irony. I abhor aubergine.

As for the plants that I had thrust straight into the earth of the paddock, the only things doing even middlingly well are the potatoes.

If there was a contest to see which crop has failed the most dismally, I would have to say that the corn might well take that title.

Of the thousand corn seeds planted, about 10 per cent have grown, which was what I had thought I would be happy with. But then their growth slowed, and now appears to have stopped altogether. There is no corn plant taller than about 60 centimetres. They should have been well over my head at this point, and although I am not a tall man, this apparently is asking too much of them.

Never has a field of corn looked so pathetic.

On the positive side, the corn appears to be remarkably resilient to wind damage. I have returned to Pru's at Weathersfield Organics to pick up some more seedlings, and her formerly majestic field of corn has been decimated by the wind, whereas mine has survived unscathed.

I need more seedlings because I am trying to master the art of succession planting. As Pru explains, the main mistake that people make is that they buy a whole lot of seedlings at once and fill their gardens with them. The sheer number of seedlings takes a lot of looking after, and there is no succession planting.

The last thing you want, she explains, is for everything suddenly to be ready all at once, and there is nothing available to come on stream once what you have used has done its dash.

Pru is wrong, however. I have discovered that the last thing you want, rather than having a super-abundance of crops all at once, is for nothing to grow in the first place, or for it to grow for a while and then die.

You would have thought that if the paddock could grow such great pasture, then surely it could grow vegetables. Clearly something is going seriously wrong. I need help.

I had met Betsy Kettle at her homestead some time previously. Betsy is a permaculture teacher, and if anyone can help me, it is Betsy. The last time I saw her, we had sat alongside her flourishing, intensively planted raised-bed gardens, and discussed the current state of the food crisis as she saw it.

Her simple abundant gardens are foremost in my thoughts as I wait for her to arrive to analyse my situation.

Betsy strolls around the field gardens, bends to look at the tattered remnants of the plants that remain, and delivers her verdict: they are all suffering from severe nutrient deficiency.

The problem, it seems, is that I am trying to grow things in dirt, not soil. The supermarket of the soil is empty.

Oh. And here I am, thinking it is my fault.

While it is good to find out that this is the case, it also provides a microcosmic metaphor for a larger problem: the sustainability of the country's soil — something that is Betsy's passion.

As Betsy explains, soil is the basis of our economic system in New Zealand. It produces our dairy and sheep products, and our fruit and vegetables. Keeping the soil healthy and nutrient rich will be the biggest problem we have in the future, especially with the end of cheap artificial fertilisers.

'Right now, we're putting our valuable organic matter in the landfill,' she tells me, 'when what we should be doing is creating a cyclic system, where food scraps and organic matter from the city is composted, returned to the rural land, grows the food, and that food then goes back to the city.'

It is timely advice. The newspapers are full of headlines concerning the exponentially increasing price of fertiliser, which is adding tens of thousands of dollars to farmers' expenses, if they can continue to afford it at all. Of course, at this stage, I have still to master compost, and tell Betsy so.

'Well,' she says, 'you have within yourself just what these plants need. You get rid of litres of the best source of nitrogen and phosphorus and potassium every day.'

'I do?'

'What you should use is what I like to call man-tea.'

'Sorry?'

'Simply collect it, water it down or you'll burn the leaves, and use it on your gardens.'

'You mean . . . ?'

'Yes,' she says with a smile. 'And best of all, it's free.'

Blissfully unaware of the ramifications that this conversation will lead to, I sit in the sun with Betsy, shooting the breeze.

'The problem', she says, surveying my domain, 'is that none of this is really that difficult. Everyone should have a garden, even if it's just to grow some herbs and lettuce, but people claim there's no time. Instead of relaxing in the garden, they'll spend an hour at the gym, or flop down and watch some rubbish on TV.'

'Of course, some of the rubbish on TV can be quite good,' I say, mindful of my situation.

'Present company excluded.' She laughs. 'But gardening and cooking shouldn't be considered chores — they should be joyous affairs. Somewhere along the way we've lost the knowledge and the love of them.'

I can understand her sentiments. The failure of my trial beds has caused me a little disillusionment, so with my dinners at stake I decide that I will try a new tack.

One of Betsy's permacultural teachings is that those parts of the garden that are used frequently should be planted in places close to the house, along well-travelled paths, so that they are always in one's line of vision. That way it's easy to recognise any developing problems, which can then be nipped in the bud.

In the spirit of Betsy's advice, I decide that I will create some new gardens in the area directly in front of my caravan, sandwiched between the cookhouse and The Chapel.

I also decide that, as these are to become the main source of my produce, I will invest in some decent soil. Dirt has failed me, and I don't have the time to create my own soil through composting.

Money may not be able to buy me love, but it can buy me some lovely soil, and it amazes me how many bags of rich store-bought soil I can fit into the back of the Land Rover. With its tail sagging down on the suspension, I crawl home with my land-bags.

Betsy is also a strong proponent of 'urban agriculture' which, amongst other things, involves the setting aside of public land for the production of food. The urban agriculturist argues that rather than simply investing in more sports fields and huge mown parks, councils and local communities need to put some of the land into gardens that produce food. Instead of more ornamental trees, fruit trees can be planted. Space should not be squandered.

I have my own related issues. After the failure of the trial gardens, I'm adamant that I will reduce the amount of wasted space to the smallest area possible. The vast area of land enclosed (barely) by my fence is unproductive to me.

Also, as I rail against the vast acreages of manicured grass that are swallowing huge tracts of formerly productive farmland, I decide to do away with the lawn.

There is still the issue of how I keep the crops protected from the hungry mouths of the goats, cows, sheep and chickens. The answer is simple. I lay out three new box gardens. The first, and largest, is protected by some swimming pool fencing I have borrowed off Demolition Pete. The second simply has a slightly sturdier net structure placed over it than the system I employed in the field trials.

The third is my pride and joy.

Lying near the garden shed in the old homestead gardens are the remains of some kind of wheeled cabinet. The wheels are of the castor variety, but their ability to swivel has been impeded by rust.

I have a plan.

I build a large boxed garden with some of the timber that Pete has salvaged. Along the top of the timber I lay a track made from pieces of the ever-useful shelving system to form rudimentary U-shaped rails. It is in these tracks that the castors run.

The wheels sit underneath a large rectangular cage, and allow it to slide on and off the garden, allowing me access while rendering it impervious to stock.

It also provides a rugged frame upon which I fix shade cloth to keep the worst of the weather off the plants.

It is, in my humble opinion, a triumph. I have succeeded in totally protecting the gardens planted in the middle of the paddock from the predations of animals.

At least, I do so in theory. After all, stock can't be blamed for nibbling tender greens if the enclosures are left open through carelessness, can they?

No. They can't.

Whilst planting these gardens I try to remember to leave the little plant tags next to the plants in the forlorn hope that I will be able to identify them at a later date.

I'm sure there will be people who are aghast that after months of gardening I still can't identify everything I have planted.

I know. It shocks me too.

Having the gardens right outside the caravan, so close that I literally fall over them — as suggested by Betsy Kettle as a central tenet of permaculture — is magnificent. It's difficult to ignore what is happening in them (but, as I have proved on many occasions, not impossible).

While the chickens form an effective barrier around the gardens by preventing the approach of slugs and snails, there is a more devious enemy: the white butterfly. It flitters by on its meandering trajectory and lays up to 400 eggs at a time in my crop. These hatch into tiny caterpillars, and within a short time, if I do nothing, the tiny caterpillar will be a much bigger caterpillar courtesy of my food stocks.

Having successfully managed to get crops to grow, I am not keen to share them with this predator.

I was always a little squeamish about caterpillars, but if there is one thing I have learnt it's this: there can be no mercy for those that will try to starve me. I'm now fearless in plucking the plump green caterpillars off the brassicas (the cabbages, broccoli and kale) and placing them in the increasingly ticklish pile in my other hand.

The chickens love the little brassica-filled caterpillars that I feed them in the early evening.

All I need to do now is maintain the nutrient base of the soil.

Hence the watering can debacle.

I hadn't realised that being a human fertiliser factory — or putting the man into manure — would prompt so many issues and quandaries and complications. The questions keep me awake at night.

What is the maximum output that I can produce in a 24-hour period? If I drink more water to increase production, am I in danger of flushing whatever nutrients are in my system straight through, so that the plants prosper at my expense? Should I take vitamin supplements, and if so, what would be the best? Would drinking more beer help?

And then there are the practicalities.

Urinating into a watering can is an endeavour that sounds easier (though no less horrifying) than it essentially is. The hole in the top of the watering can is generally not as large as one supposes, and it isn't for several damply frustrating days that I realise that the best option isn't to perfect my aim, but merely to hold the watering can closer to the source.

I know what you're thinking.

I never thought I would have become an expert in this sort of thing either.

All I can hope of this ordeal is that the knowledge I have gained is of some use to you.

Now all I have to do is convince myself, and any potential dinner guests, that the lettuce has been washed really well.

22

Making an ass of oneself

I never imagined I would actually commit the sin of coveting someone's ass. However, there is a first time for everything. As the other floats for the Helensville Christmas Parade begin to take shape around me, I begin to suspect that in lieu of an ass I should have borrowed some children, as they seem to be the predominant float accessory.

I have only ever been to one Christmas parade before. I shan't name the town, as its parade, while not lacking in community spirit, had been a little lacklustre. To a large degree I was the parade (well, me and some kids on their bikes), so I was unaware of quite how seriously some people take their Christmas parade floats.

I have decided to enter the Helensville parade with a float that has a 'Sustainable Christmas' theme. Note that when I say that I have decided to enter the parade, what I really mean is that Jane has informed me that I will be entered in the parade and that some kind of sustainable living theme would be appreciated. By appreciated, I deduce she means mandatory.

The evening before the big day, I laboriously shovel rotting sawdust into my trailer as a base upon which I will create my idyllic sustainable garden scene.

Despite the physical exertion required to fill the trailer, the shovelling takes considerably less time to complete than does the tying down of the cover, a procedure that takes far longer than it should due to the exceedingly elaborate nature of my cover-fastening technique.

A quick raid on my garden the following morning sees it pillaged, as I select some of the better examples of my produce to plant in the trailer. This presents something of a challenge, as I want to have good produce on display while retaining the best produce for the event that will be My Upcoming Christmas Family Lunch Fiasco.

This is yet another of Jane's good ideas that I have somehow managed to mangle.

She'd thought, quite rightly, that it would be a good idea if I were to invite my family for Christmas lunch, and to feed them a wonderful meal that I had grown myself.

So I invited them.

What I hadn't considered was that my family is now a lot bigger than it once was. No longer is it simply my parents, brother and sister. My siblings are now both married and have whelped large broods of brats. There is also my grandfather and various in-laws.

By the time I realise what is happening, I am startled to discover that I will be catering for 17.

And me.

That's 18.

Oh dear.

As I fossick for enough produce to take to the parade, I have grave doubts that I have grown enough food even to get me through to Christmas, let alone feed my entire clan.

Once the produce has been selected, it is a simple matter of packing the Land Rover to the gunwales with all manner of bits and pieces and tools that I think I might need when constructing a work of art as monumentally grand as the float I am planning, before heading off.

At the assembly area in Helensville, there is a healthy sense of bonhomie amongst the participants. The end of the year is nigh, and the horrors of the last few shopping days till Christmas are still a wee way off. It is as if we are contently drifting through the pre-Christmas phoney war.

There are floats with angels and floats with pigs, and some floats with pigs dressed as angels. Someone has even dolled their youngsters up as Mary and Joseph and given them a real donkey.

How can I be expected to compete with that? All I have is a slightly-too-tight, terry towelling, shorty Santa suit, and a red elf hat with a white pompom on the end of a spring attached to it.

With no time to lose, I set about recreating a veritable mini-Eden on my trailer. Into the sawdust I replant a lettuce, which by now is looking somewhat limp, followed by a potato plant that looks even worse. Not even a good dousing with water will perk them up.

I have also brought a small taxidermied goat from my personal collection of taxidermied animals that adorn the confines of my now infequently visited home in Henderson. I plan to stand it on one of my bonsai lawns at the base of my chair as a companion. I believe a small animal will not only contribute to the sustainability theme, but will also be charmingly reminiscent of a biblical nativity scene.

Quite what role I represent in the nativity scene is a moot point. Given that my sheep have disappeared, I can't claim to be a shepherd, and if you've read everything up to this point, you'll agree that I can't really argue that I am a wise man, and the role of the ass has already been taken.

Around the outside of the float I have strung strings of silver tinsel from battens I have fastened to the corners of the trailer, creating the appearance of a fence of festive barbed wire.

It is only once I step back to inspect my handiwork that I realise two things.

The first is that the trailer is so small that it appears ludicrous behind the vast bulk of the Land Rover. I have the only float in the parade that is less than half the size of the vehicle towing it.

Secondly, I discover that I have industriously fenced the entire trailer, and thus have left no convenient way to get back into it. This might not have been so bad had the judges not chosen this moment to arrive, prompting me to crawl through the tinsel wire in an undignified manner as they stand regarding my efforts with what I can only interpret as a look of stunned awe spreading across their faces.

After some po-faced scrutineering of the entire parade, the judges confer, and declare the winner to be a large truck with a Robin Hood theme, which features a live pig, albeit not one dressed as an angel.

I console myself with the fact that winning isn't really what the day is about, and that all I need to reward my efforts are the cheery smiles of the huddled masses of Helensville. This works until I realise that the judges aren't giving out certificates that celebrate participation.

Scandalous.

I have recruited Demolition Pete to drive the Land Rover, and as the parade makes its way out of the assembly area and onto the road, I become concerned that he is going to have a tough time of it. The Land Rover's clutch is a bit stiff at the best of times, and as the parade keeps coming to a halt for no apparent reason, it is clear that he is in for a good leg-based workout. All I hope is that by the end of the parade I will still have a clutch.

When the convoy isn't mysteriously halting, it is travelling along at such a great clip that the camera crew have trouble keeping up. I rejoice in the fact that I now have the chance to sit back and watch them sweat and struggle in the sun.

Winding our way up the main street of Helensville, I become aware that I am getting a very cheery and vocal response from the more disreputable-looking elements of the crowd. Never have so many shady characters nodded and winked at me, and I'm stating this from the position of someone who is normally a magnet to the shadiest of characters.

Perhaps they are concurring with my message of sustainability, I think, until it occurs to me that the public might actually be misinterpreting the many innocent mantras that I have attached to the float, notably the 'Have a merry green Xmas', and the 'Grow your own green Xmas' signs.

Am I simply being paranoid? Maybe they really are saluting the audacity of my float-based brilliance. Surely people couldn't think that what I mean is something a little more subversive than the mere growing of a sustainable lunch? And if they do think that, then is it any wonder the judges looked at me the way they did?

There is nothing for it but to keep smiling and waving.

A lot of people ask how difficult it would be to be the Queen, when all you really need to do all day is wave at people. I can now confirm that waving at people can be very exhausting, and I have employed a technique that requires as little effort as possible, based on one I have seen the Queen use — the raised, slightly cupped hand, on minimal rotation method.

If anyone knows how to wave a wave, it's her.

As the sounds of the adoring crowd fill my ears, the sun beats down on my legs, rendering them a little pinker, and the smell of an overworked clutch fills my nostrils. Children lining the roadsides make the most of suddenly being allowed to take sweets from strangers. I hope there won't be a repeat of the incident where children from Kaikohe mugged Santa.

I wave at the camera crew as they huff past me in order to get another shot.

As the parade halts again for reasons apparent to no one, and I'm Ho! Ho! Ho!-ing in the general direction of a group of children, the crowd suddenly starts laughing and pointing at me.

I always feel uncomfortable when this happens, and I have no idea why, until I turn around to see the last of my float's lettuce being eaten by the donkey.

'Merry Christmas,' I say.

23

Turkling

As my nieces and nephews, clad in their Christmas Day best, peer into the plastic pet carrier at the quivering baby turkey, they ask: 'Where's its mum?'

'She's over there under that tea towel,' I say. 'We're having her for lunch.'

While some children may have been shocked by this, my four-year-old nephew, Rory, points at the tiny turkey in the cage and says: 'Can you grow this one all big so we can eat it next Christmas?'

'We sure can, buddy,' I reply.

If I had known that the turkey that has given up its life for Christmas lunch had babies stumbling around in the long grass at its feet, I would never have shot it. As it was, I didn't know, and within the space of a few moments, the baby turkeys were orphaned in order to feed my family on Christmas Day.

There had been three babies. One ran away and couldn't be caught; the second suffered collateral damage from the shotgun scatter and had to be put down; but the third showed some fighting spirit, and Jane decided to take it home and put it in her hot water cupboard as I had nowhere warm to house it that was anywhere near comfortable, or appropriate.

The reason I had to go turkey shooting is that I somewhat underestimated how much food I would need to feed the masses that have turned up to help me celebrate the Christmas experience. The turkey and her children have been casualties of my Good Host Anxiety.

It is a few days before the fated Christmas lunch, and as most of the rest of the country flees to the supermarkets and the malls, falling over each other in a consumer-oriented buying spree, I am strolling around a sun-drenched meadow with Shawn, looking to supplement my Christmas lunch with a little free-range turkey.

I don't know what you would pay for an organic turkey in the supermarket, but the three I shoot are free. The only food miles I use are the ones I walk to get to the birds.

There are no food miles in the main course either. It is more a case of food metres, as the source is my own paddock.

We are to have lamb, which is good for my guests, but not quite so good for the sheep. This is a dish that will require no trip to the butchers to stand in front of the freezers, oohing and aahing at the price of meat. I have lambs, and all I need to do is hustle them from the pasture to my plate.

The rules of the home-kill industry are no doubt well intentioned, but are routinely ignored, if they are in fact known at all. Under the regulations of the Animal Products Act, those who can have a beast slaughtered are those who have 'actively engaged in the day-to-day maintenance of the animal' and the meat 'must only be used for consumption by the direct family, or household, of the person, or a farm employee and their family or household'.

So, those of you who receive someone else's home-kill are breaking the law.

This has to be one of the least followed laws in the country.

However, as we are filming it all for TV, it means I need to get a professional in, and so within a short while the home-kill folks show up. He is tall and thin, and she isn't. They arrive in a large 4WD, towing a big trailer containing the carcasses of other people's stock that has already been dispatched that day.

This was the way I had always known meat to be collected. As a child we would often help to round up the selected cow, generally a young heifer that was sleek of flank and shiny of eye.

Mother would take us down to watch as they shot the cow, hung it up at the back of the truck, then speedily skinned and gutted it. This would all be done next to a hole that we would have to dig, into which would be put the ample innards. We would take the liver, kidneys and heart home to cut up for cat food.

As a way of making polite chitchat with the home-kill folk, I ask what else they have killed recently.

'We did some ostriches a while ago. They're difficult — small heads, ya see,' he says.

Within a very short space of time, I go from pointing out the unfortunate sheep to holding a large bucket of offal, skin, liver, kidneys and heart, all darkly purple and warm, and the home-kill folk are disappearing back up the driveway to continue on their killing spree.

As the Christmas season is a time when a man needs all his wits and strength to survive, the following morning I slice and dice the liver, kidneys and heart, and coat them in a little flour, before gently frying them along with a couple of eggs that I collected that morning, over the roaring potbelly in my cookhouse.

Sitting in the warmth of the early morning sun, I look up the hill to where the other sheep are standing, and hope that they too will taste as good. I need my strength as I have a cow to milk.

As a child, Christmas was entirely based around cows. They rightly took priority; after all, they were the ones who were paying for it. They had to be milked before we could open our presents. If we were to leave the farm for Christmas lunch, we had to be back in time to milk them again. Late lunches were the worst, as all you wanted to do was sleep off the gluttony, but that wasn't an option with the hundreds of cows waiting.

Judy's bountiful udder will once again provide me with cheese. This time though, I don't have to contend with the hungry mouths of Campbell and Sainsbury, and for my efforts receive a good pail of milk. This is taken back to The Chapel, where my big shiny new barbecue proves adept as a vehicle for the creation of cheese. I am not sure how many people have created cheese on a barbecue, but I can highly recommend it. Using the simple techniques that Sabine has shown me, I soon have the feta on the go.

My big shiny new barbecue came about because I realised that if I was to have any chance of feeding my family on Christmas Day, I would have to take some drastic action and invest heavily in infrastructure.

I did not become the proud owner of a giant barbecue as the result of a spur-of-the-moment descent into consumerist abandon. Oh no.

I did my research before I shopped around, found one I liked, and bought it on the spot. It was the top of the middle of the range: a four-burner monster with side element, hood, cupboards and thermometer. I particularly liked the knobs, which looked as if they were designed by someone whose last project had been to design the knobs of a stereo, circa 1983.

It looked huge, especially in its box.

I brought it back to the farm, unpacked it, and began to assemble it. Then I thought I should have a quick squizz at the handbook, just on the off-chance that I was doing it wrong, and because it didn't seem to want to go together very easily. I realised that the reason it didn't seem to want to go together easily was that it appeared to have been damaged in transit.

I disassembled what I had assembled up to that point, and attempted to repack it.

It's one of the strange anomalies associated with returning damaged goods that no matter how hard you try, they can never be refitted back into their original packaging. By the time you have them packed neatly back in the box, packed a second box with everything that won't fit in the first box, and have located the original receipt, the warranty has generally expired.

With most of it back where it belonged, I returned it to the shop where my worries about a confrontation were allayed by the salesman, who simply looked at it with a resigned expression that said, 'Wow, another one, what a surprise', and replaced it without question.

Without doubt, this entire affair has increased the size of my carbon footprint to clown-shoe-like proportions.

By the time I finally had it assembled, the floor of The Chapel was lost under a mountain of packaging, and the late afternoon sun was glinting off the barbecue's metallic surfaces.

I stared at its potential. I now felt like I could entertain.

So long as I remembered to fill the gas bottle.

While the feta is maturing, I have time to deal with a more pressing crisis. I intend to serve boneless lamb and turkey, both roasted and smoked. But I can't really serve only meat for lunch. I'll need a range of side dishes and desserts. All I have to do is figure out quite what those sides and desserts are going to be, based on the very limited amount of produce I have.

Once again, I need a wise woman, and lo, like a mystical yuletide vision, she appears in a late-model red Mercedes and snug jeans.

Annabel Langbein is the answer to a number of needs I have. Trained as a horticulturalist, she trapped possums for a living before becoming the bestselling author of a number of food books. What more could I ask for?

She is here to restore hope where there is only the fear of my inability to feed my family.

'Welcome to my Christmas catastrophe,' I say, indicating my scruffy garden.

'It's not looking that flash,' she agrees, in what I consider to be a masterly understatement.

As we stroll through the gardens, Annabel laughs at my beans, is dumbfounded at the poor state of my rhubarb, and can't help noticing the lacelike quality of my silverbeet.

'Lucky we're not relying on you to feed the world,' she says.

Thanks, Annabel.

I'm feeling a little demoralised. It isn't supposed to be like this. But as we walk around she points at edible things where all I see is foliage. It is astonishing to walk around the gardens, that only moments before I had been looking at with abject despair, with someone who knows what things are, pointing out edible species that I hadn't even realised were growing there.

'You need me, Radar', she says.

'That's becoming increasingly obvious,' I reply. 'Wish you could come and live with me.'

Kneeling by the potatoes, she asks if I know what bandicooting is.

I do not.

She buries her hands carefully in the soil under the plant and they emerge with a plump new potato.

'When you're harvesting your potatoes,' she says, 'you need to get right in under them and take a few potatoes out, but you don't want the plant to know or it will get grumpy and die.'

'I had been hoping to have enough potatoes to make vodka,' I comment.

'I could help with that too.'

She is the perfect woman.

I scamper along behind her as she bustles though my acres. She points out a stand of tall plants.

'Look at those, they're Jerusalem artichokes,' she says, once again falling to her knees and wrenching them out of the ground to reveal a knobbly tuber. 'You just need to scrub them off and roast them with the lamb. You have to be careful though — they can make you a little gassy.'

I have been blind to the wonders that are being produced before my very eyes. Not only can Annabel identify them, but she suggests ways of preparing them. She is a ray of sunshine in an overcast garden.

I take notes for making salsa, lemonade and dessert.

By the time she's ready to leave, I am quietly confident that all will be well, just as long as I don't poison anyone.

'You have a feast happening,' she reassures me. 'I'll be on my way then.'

As she heads off up the driveway, I stared at the little notebook in my hand, trying to decipher what it is that I have written.

It's the morning of the big day, and I am disappointed to note that Santa hasn't stuffed my stocking full of any joy.

I have rather a lot to do before my family arrive, as the place needs to be spick-and-span for them. There's nothing like a big pre-function tidy-up, and I want the place to be super-spruced, although I realise halfway through the morning that no matter how tidy I make it, Mother will no doubt know how untidy it was before her arrival. Mothers know everything. It doesn't help that the process will eventually unfold before her eyes on television.

I have to finish the chicken coop and tidy the gardens and remove certain organic matter the cows have left under the totara, which is where we shall be dining, then round up the vegetables and bake the beer bread and slather the lamb with oil and rosemary and rub the turkey breasts with honey so they can be smoked and prepare some salads and put out the tables and chairs and make sure the plates and glasses and cutlery are in place and make table decorations and prepare the rounded-up vegetables and check the bread is cooked and get the cheese and make the hors d'oeuvres and ensure that everyone has a cup of tea when they arrive, especially Mother and Grandad, and make sure that all the booze is chilling.

The more attentive readers will be aware that not even in a month of Sundays can I achieve this on my own, not even if I sell my soul (even supposing I can negotiate the rights to it back from Jane).

I will be eternally grateful to Jane and Deirdre and Ruth for doing all the things I haven't the time to do, but which through the magic of television it appears I did. Women are so damned useful.

Within short order the turkey is roasting with the lamb in the barbecue and the potatoes are on.

I've done as Annabel suggested and bandicooted my way around the plants. I slid my hands underneath, and there was potato after potato. What a harvest. It is a gardening triumph.

It's funny how the little things can seem to be the biggest victories, and the potatoes are a decisive battle with the land that I have won. I know now that I can grow things.

I decline to do anything with the abundance of Jerusalem artichokes as I feel my family will never forgive me if I send them all home on a long car journey suffering from the ill wind of the little root vegetable. That is not the particular summer breeze they are after.

The family arrive early — of course — and I wrangle the kids in to help, and pretty much on schedule, lunch is served.

By sheer coincidence, this is also the day that we have calculated the broody chicken will be hatching the eggs that we have placed under her. We have no roosters, and so fertilised eggs have been purchased from TradeMe and added to the chicken's little cache of sadly barren eggs.

The plan is that when the chicks hatch, we will slide the little orphaned turkey in underneath the hen and see if she will adopt it. I've been referring to it as a turkling because I have no idea what the correct term for a baby turkey actually is. Jane simply calls it Christmas, which seems appropriate.

We will be able to tell pretty soon whether Christmas has a fighting chance in the world, as research indicates the hen will either tolerate it, or peck vigorously at its head until it dies.

With the turkling's mother in a better place (my family's stomachs), I cup it in my hands and carry it to the nest the broody chicken has created in the tall grass. Cautiously approaching the chicken with my young nephew Jake, we place Christmas in the crook of her wing, and wait to see what happens.

Without hesitation she simply wraps her wing around Christmas, drawing him into the warmth and safety of her feathery bosom.

It is a wonderfully touching Christmas moment.

24

Roosters have little to crow about

I love chickens, but not as much as I love chicken. The question therefore is not whether I will kill some chickens, but when and how. The time has arrived when that 'when' is now, and the 'how' is to be with an axe.

I have used that mighty repository of human endeavour and knowledge known as the Internet to research the various options for dispatching chickens, and concluded that the time-worn method of removing the chicken's head from its body with a sharp axe appears to be the quickest and the most humane.

It certainly seems far less risky than the methods that advocate wrenching its neck, or taking a thin knife and stabbing it through the roof of its mouth and into its brain. Some people even suggest gassing them, but who has a handy canister of anti-chicken gas lying around?

The most outlandish proposition suggested in all seriousness by several websites is that before I attempt to kill the bird, I should calm it down by hypnotising it.

Selecting which bird is to be for the chop-chop isn't overly taxing, but it does illustrate the inherent sexual inequality of the farm. More often than not, it is the male of the species who is slain, as they are of less use than the productive females. Hens lay eggs, and cows bear calves and milk, but their respective menfolk do little but fight and attempt to make love.

Or fight while attempting to make love.

And when they are not doing either of those two things, they are generally lying around doing very little.

I've had my eyes on the chickens from an early stage, trying to figure out which of them are male. As they have matured, it has become fairly clear which ones they are.

As pre-pubescent poultry, they first began experimenting with their crowing, and the ones whose voices were breaking didn't quite manage to add a 'doodle doo' to a strangled 'cawkle'. They were blissfully unaware that in doing so they were dead chickens squawking.

It is only a question of which of the roosters it is to be.

Two of the five roosters have disqualified themselves from the selection due to general oddness. One has blossomed into a small but paunchy, long-necked, grey bird, who struts around with the demeanour of a young Napoleon.

When he was a young chicken, I had thought that he was deformed, for it seemed that his head was on back to front. He would try to walk, but as his eyes were facing backwards, and his legs were moving forwards, all he managed to do was look positively perplexed for a few seconds, then fall over. How could I not have been smitten with the little guy?

The other bird that has saved himself from the block is a bigger, more conventional-looking rooster, but he has feathers growing all the way down his legs and out along his claws. They remind me of the sideburns that Hugh Jackman wore as Wolverine in the film *X-Men*. The name has stuck.

These two chickens have captivated me, for I realised at a young age that the creature that could draw attention to itself for being different, or quirky, could, if it was lucky, have a more charmed existence.

When I first arrived at boarding school, there may have been other boys who were as frail and as sensitive as me, but I alone was ginger, freckled and bespectacled. I certainly didn't need a sign that said 'I'm different from the other boys' to draw unwanted attention to myself.

As with the chickens, the fact that I was unlike the other boys had its advantages in surviving the killing fields of the dorm rooms. An older boy, generally considered to be the toughest in the school, decided that he would take me under his Old Spice-scented wing. This ensured that I was his, and his alone, to torment, or protect, as was his whim, and it was often his whim to do both simultaneously. He was like the older brother I never had.

As my other male chickens are relatively unexceptional, I have formed no attachment to them. Whenever I look at them, all I really see are different portions of chicken, prepared in any number of ways.

Which of them I select for my inaugural home-grown chicken dinner will come down to whichever of them I can catch first.

With the hour of the wretched rooster's demise looming, I have honed my tomahawk to a razor-sharp edge, and as I trudge to the chicken coop, I can't help but think of a line from *Macbeth*: 'If it were done when 'tis done, then 'twere well it were done quickly.'

I have chosen to feed my brood later in the day, so that now, at the appointed hour, I can entice them all into the coop. With their heads in the feed tray, I pounce.

It always amazes me that when held upside down, the chicken will cease all resistance. It's as if the pressure of the blood on their tiny brains is too much, and they simply cannot process why the world has suddenly become so disorienting.

The rest of the flock seem totally unperturbed by their mate's sudden disappearance. It's as if he was never there.

The chicken's inability to function while inverted means the whole process is conducted with a quiet dignity, or at least as much dignity as executing a bird condemned by his sex to an early axing can have.

I carry him to the stump, lower him down so that his head and neck rest on the wood, and promptly separate said head from the neck with a single swift thwack of the axe.

As the legs of the bird convulse spasmodically and his wings engage in a frenzy of flapping, flipping traces of blood across my face as it squirts from his neck, I turn to Jane and say, 'I suspect we won't see that bit on the telly either.'

'No,' she agrees, 'we most certainly will not.'

Plucked, he is a little leaner than I had imagined he might be, but it comes as no surprise. My brood are an energetic and freely ranging bunch whose territory encompasses a broad swath of paddock and section.

While all the exercise means they are a little leaner, I do encourage this, as it means that while they roam around, they are fulfilling their roles as a feathery buffer protecting my crops from the incursions of snaily and sluggy predators.

Of all the meals that I will prepare on the farm, this is the one I am anticipating the most. I've known from the moment I arrived and realised I would be eating my own chicken, exactly how I will prepare it.

I intend to butterfly my chicken, and roast him. The process of butterflying a chicken — essentially cutting out the spine and opening up the bird so that it looks like a butterfly — is a technique I have picked up from the Internet. It's simple and quick, and results in a lovely, easy-to-cook piece of chicken that is also easy to divide when the time comes to serve it.

I've even waited until I've been able to construct a special oven for the purpose from an underrated, totally sustainable source — mud.

Mud is fabulously versatile. It's an entirely renewable resource that can be used to create magnificent structures. I intend to use it to satisfy a grudge I harbour against television chef Jamie Oliver.

One evening, while watching an episode of his *Jamie at Home* series, I watched in horror as he whipped up a wonderful pizza and some bread, and then smugly popped them into his large outdoor pizza oven to cook.

I was aghast. 'Who in God's name can afford to plonk a pizza oven in their back yard?' I shrieked at the television. From Jamie's smirk it was clear that he could.

As it turns out from my research, anyone can have a pizza oven for very little cost and only a modicum of effort. All you need is a little mud, some river sand, some straw, and a back yard. You could in theory site one of these on the balcony of an apartment, but I suspect the neighbours might get a little antsy. No doubt they could be won over with some of the delicious food that would emerge from it.

The other prerequisites are a small amount of time, and having no qualms about getting dirty.

While in Mali I had the great pleasure of encountering the world's largest mud building, the Great Mosque of Djenné. It is a massive monument to mud in a country where mud seems to be the main building material. So large is it that up to 3000 people can be accommodated in its main prayer hall. No doubt they all pray occasionally that the mosque will not fall down, or wash away.

For most of the year, their mud buildings are ideal structures, providing cool respite from the scorching Saharan heat during the day, while remaining warm at night. When the rainy season arrives, the weather seems to do its damnedest to wash their houses away. In a not-so-vicious circle, this simply produces more mud, which they can then use to patch up their houses.

Prior to my visit to Mali, the closest I had come to adobe was the Photoshop program on my computer. I left the country a convert to its versatility.

It's an overcast morning and I am waiting for Wolfgang to arrive. Once more his battered Volvo rattles into the paddock, his trailer laden with several buckets of mud. This isn't any fancy mud, if such a thing exists. It is simply a mixture of ordinary clay, sand and water.

When I ask about the secret to combining the right amounts of each ingredient, he explains that in order to determine the correct amounts to produce your workable mud, simply blend a small amount of what you think is the right combination, roll it into a ball, and if it is dry by the next morning and isn't falling apart, then you have it pretty right.

Despite my initial scepticism as to Wolfgang's choice of locating the site for the oven directly under the totara tree, he has promised me that I will not burn the tree down. I certainly hope he is right. That would be embarrassing.

So, we begin.

Once again, the discarded concrete fence posts come in handy as we use a large stack of them to create a waist-high base, on top of which we rest four large concrete paving stones. This forms a sturdy surface to place a layer of ordinary clay bricks. You can use firebricks, but I haven't managed to scrounge any of these.

The bricks are then cemented into place with a little of the mud. Using a 40-centimetre stick, we measure the radius of the circle we etch onto the base, noting where the centre is. The stick is then stood upright in the centre, and sand is piled around it to create a dome, as high as the radius is wide. Judicious use of water keeps the sand damp and in place.

Covering the sand with wet newspaper, I note that many of the pages are advertising real estate, mostly beachfront properties. How many of these overpriced retreats would have a mud oven? I wonder. It is immensely satisfying to bury these turgid pages of aspirational real estate porn under a heavy caking of mud slurry (to a depth of about 10 centimetres if you are reading this for instructions, and not pleasure).

On top of this goes a thick mixture of straw and mud. This needs to be of a consistency whereby the straw is just held in place by the mud. This 10–15-centimetre layer creates the key to the stove, a thick thermodynamic layer designed to contain small air pockets that will be heated, and which will retain this heat for a long period of time.

Another 10-centimetre layer of the original slurry covers this, which we smooth off, and then all that is left to do is to create the door by cutting some of the mud out of the wall. The height of the door is one of the most crucial elements in the process. It should be about 66 per cent of the height of the interior of the dome.

By the end of a pleasant morning, we are finished. Wishing me luck, Wolfgang repacks the Volvo with Teutonic efficiency and departs.

After allowing it to dry for a few days, I delicately remove the sand mould from the interior, fearful that the mud inside will collapse and I will then have to start all over again.

Try as I might, I cannot get the fire to catch in the centre of the oven. I light a little kindling in the gaping maw of the oven, get it burning and push it delicately inside, where it immediately dies down, splutters a little smoke, and dies.

Clearly it lacks oxygen, so I carefully shave the top of the door off until it allows enough air in to keep the fire going.

The great thing about working with mud is that if you remove too much, you just need to splatter a little more on. I enjoy working with steel for the same reason. You can often fix it with a little judicious welding.

After a few days of lighting increasingly larger fires, it is time to cook something.

Heating the oven is straightforward: light a fire inside, keep it burning at a fair clip for a few hours, then scrape out any coals or embers, and insert food.

Wolfgang has warned me that the first time anyone ever tries to cook something in one of these ovens, they invariably burn it. I can see why. The heat inside the oven is phenomenal. I insert my oven thermometer into its furnace-like interior and take a reading. The result? One melted thermometer.

In light of this, I christen my mound of mud The Kiln.

Before I sacrifice one of the roosters to its fiery depths, I experiment with a goat casserole. Into my cast-iron camp oven go a little haunch of goat, some new potatoes, carrots, a handful of herbs, and a generous dollop of red wine. The camp oven goes into the kiln, and the entrance is sealed.

Some people fashion might little doors for their oven to trap the heat in, but I prefer to use a series of bricks. They are heavy, get quite hot, and never seem to fit together the same way twice, but sealing up the hole with the cumbersome bricks gives me a sense of satisfaction.

Within seven short minutes, the camp oven appears to be glowing faintly white, so I think it prudent to withdraw it and inspect the contents. The wine has all but disappeared, and the meal appears cooked.

Crikey.

While it could have done with a longer, slower cooking, it is delicious. Discounting the hours needed to heat it with fire, the kiln appears to be able to cook food faster than a microwave.

It is clear I will have to learn the subtleties of the beast, but I am rapt. I have cooked a meal in mud. How sustainable of little old me.

Over the following months, come rain or shine, the kiln sits nestled squatly under the protective canopy of the tree, and from the blistering heat of its interior emerges all manner of bread, pies, pizzas, roasts, baked goods and casseroles.

And now it is time to test it on a roast chicken.

After I slide the body of the bird into its depths, along with some compost-sprouted potatoes I have transformed into spicy wedges, I have time to relax and contemplate the process.

When you take into account the length of time and unpleasant nature of killing, plucking and gutting the chicken, and adding in the cost of its feed, and the need to have someone feed them should you be away, you have to wonder why it is worth it, when they can be bought from the supermarket for as little as $7.

Even paying the premium for free range, or organic chickens, it's still cheaper than growing and killing your own.

My chicken may not be as plump or tender as a store-bought one, but it tastes like no chicken you will ever buy.

It just tastes chickenier.

As I pull a drumstick off the cooked bird, I know that this chicken has had a full and rich life doing all the things that chickens love to do, right up to the moment I severed its head.

Sitting under the totara tree, eating my glorified chicken and chips, I look up to see another rooster sitting contentedly on the fence, looking at me.

I wave a drumstick at him and keep eating.

25

Barry the lamb

Jane has entered me in a show. Well, not me exactly. I shan't be displayed in a cage, or leashed to a pole like some kind of carnival freak. Rather it has been decided that I am to enter some of my produce in the produce-growing competition that forms a part of the great spectacle and wholesome family fun day that is the annual Helensville A&P Show.

Oh dear.

That the good folk of Helensville and the surrounding areas will see my produce is a more fearful prospect than the somewhat more distant certainty that vastly more people will witness my inability to grow much of anything when the show is televised.

My immediate concern isn't that I will be competing against people who have been entering the competition for years, and who have honed their competitive garden skills to a fine art. I'm more concerned that there is a very real probability that the competition may force me to go hungry, as I'm unsure whether I will be able to have my vegetables returned after the event. It would be a great shame to have to enter the best of what little I have grown and then be precluded from eating it.

Still, I do have to confess to being a little excited at the prospect of seeing how my produce will shape up.

I have never really won a prize before. I was the top horticulture student in the third form at school, but given that two per cent of that exam was awarded simply for writing your name on the paper, and that some people managed to get this wrong, it is little surprise I topped the school.

As the competition is organised along lines of ruthless efficiency, I have to declare some time in advance the vegetables I am going to enter.

'I shall enter all of the categories!' I proclaim to Jane.

She looks at me with the look of sad derision I have come to know and fear, and informs me that not only have I not planted half the categories, but of those that remain there is the real likelihood that many won't survive until the day of the show.

'Oh. Well then. Umm, what can I enter?'

The entry form that is eventually submitted declares that I, Te Radar, will be throwing into the ring a watermelon, some parsley, potatoes, beans, peppers, corn and eggs, even though I guess the real credit for the latter's success will ultimately have to go to the chickens.

With each passing sleep, the show day draws closer, and I try to plump up my vegetables by adding extra cow manure slurry, and watering them with both worm and man tea as often as I can remember to. I'm not sure it is doing much good, or whether it is possible to overfeed them, but at least they are not dying.

Of more concern is that I have to take the caravan to the showgrounds so that I can stay overnight. If there is anything that will make the Land Rover even more unwieldy, it's the addition of a caravan that has the aerodynamic properties of a large concrete slab.

I am right to be worried, for as the moment for departure arrives, the weather proves to be quite dynamic. An approaching storm is being heralded by a blustery wind that is buffeting both the Land Rover and the caravan to such an extent that I am worried they will be upended.

But the A&P show waits for no person. I have produce to enter, and I can't wait for a lull in the storm.

Shouting a cursory 'Wish me luck!' to the cows, I wave them a cheery goodbye, and with a graunching of gears lumber off up the driveway, trailing a thick plume of diesel smoke.

Somehow, despite being very nearly blown off the road on several occasions as the caravan, caught by gusts of wind, fishtails behind me with all the grace of a drunken sailor returning from a rather celebratory bout of shore leave, I arrive safely at the A&P showgrounds.

I select a site in the car park near where a few other house-buses are parked, and as a light drizzle dampens the soil I'm feeling a little smug, for while others will have to get up tomorrow and struggle through the wind and rain to get to the show, I'm already comfortably in residence.

With not a moment to lose, I rush the box containing the cream of my meagre crop to the table accepting the produce entries, pay the entry fee, and return to the caravan for a much-needed nap.

Near midnight there's a knock on the window, and I pull the curtain back to reveal the wide-eyed faces of several teenage girls. They are the same troop of feisty young Scouts who came to the farm to make damper with me, and I am reminded that I still haven't embroidered them any achievement badges.

They are the reason I'm in residence at the grounds, as I have been conscripted to assist them in patrolling the site. I had been worried about what would happen should we discover any untowardness, but as the troop appears to be fuelled by sugar and the critical mass of their own unbridled adolescence, I no longer fear encountering a miscreant.

Rather, I fear for the miscreant who might be unlucky enough to encounter us.

These femmes could prove fatal.

As we traipse around the site, I conclude that the chances of encountering any scoundrels are slim, given the shrill laughter and girlish squeals emanating not only from me, but also from the girls themselves. I insist we check the food caravans, as I would be inconsolable if my long-awaited battered hotdog didn't eventuate the following day because some burglar had pilfered it in the night.

After roaming through the showgrounds for some time, shining torches at quad bikes and gazebos and tractors and fences and stables while engaging in various high jinks, I am wilting under the onslaught of the Scouts' gusto. Thankfully, our tour of duty ends, and we return from the night like grizzled veterans to the building they have commandeered as their bedchamber.

As they clamber into their sleeping bags, I bid them a hearty goodnight, and accept the offer of a refreshingly wholesome cup of Milo and a biscuit in the adjoining room with the bearded Scout leader (who is a man, as it happens). He informs me that the show patrols originally started 20 years ago after the audacious theft of a marquee, but that they have never actually encountered any problems since.

After having experienced the diligence and fearsomeness of these young women, I am not surprised.

As the long-awaited rain begins to fall in great torrents from the night sky, I retire for the evening.

The irony. After a long drought, the rain has decided to come on the very day it isn't really wanted. A great deal of money is tied up in the event, and the showgrounds have a reputation for becoming very muddy underfoot, meaning punters will not want to venture out in the atrocious weather.

Listening to its moist goodness pounding onto the roof of the caravan, I lie awake in eager anticipation of the day ahead. As a child I was never a big attendee of A&P shows, nor Easter shows, or gymkhanas for that matter, and I am keen to make up for lost time, and to literally experience all the fun of the fair.

At dawn I fling open the caravan's door to be greeted by a lacklustre, damply grey morning, and a fairground that isn't exactly shrieking 'Fun to be had here!'

The one bright light on the horizon is provided by the sight of a passing woman who has clearly not really thought about where she is venturing that day, and whose high heels are proving less than ideal footwear for the rain-sodden ground.

As she attempts to unstick one from the puggy earth into which it has impaled itself, her other heel sinks in deeper, until she is literally nailed to the ground. She looks at me in confusion, clearly unable to comprehend the situation. All I can do is smile back and offer a cheery 'Good day for it!'

It isn't though.

It is reassuring to think that if the worst happens and fleets of cars are to become stuck in the parking paddock, then as it is a showcase of all things rural, there will be no shortage of tractors to haul them out.

What there is, though, is a shortage of people.

The people who have come wear warm jackets and worried smiles. It is bucketing down, and the forecast is for continuing rain, but rural folk are made of sterner stuff, and many continue on oblivious to the rain.

I'm drawn to the Calf Club regional final that is being held in the drizzle. Calf Club, for those of you unfamiliar with the concept, is a quaint rural phenomenon where parents allow their children to rear a baby animal of the ungulate variety (calves, lambs, or a goat). The children are supposed to spend several weeks teaching the animal to obey simple oral commands, and mastering the art of leading said animal around a hessian-lined show ring on a piece of string. The children who do this the best win a ribbon.

If nothing else, it proves the old adage that you make your own fun in the country.

As a child, despite growing up on a dairy farm, I was never allowed to rear a calf. Mother was a firm believer that it was the other mothers, and the occasional father, who did much of the pre-show grooming, and she was far too busy to be tied down to this commitment.

She did, however, allow us to borrow lambs from the Wiggins family. Oh, how my siblings and I would rejoice in the rapidly wriggling tails of our lambs as they guzzled from their bottles of milk — at least before the rubber ring claimed the tails.

I was clearly not very adept at training them though, as I never seemed to win anything but the charming smile of my lamb as we frolicked in the spring sunshine.

Once Calf Club day was over, the lambs were returned to the Wiggins, where no doubt they lived a full and satisfying life, dying eventually of old age in their sleep.

One year, things were different. I had a lamb that seemed to be no ordinary lamb. I called him Barry.

Although a little on the puny side, which only served to endear him to me even more, Barry proved to be fairly adept at obeying oral commands and walking around in rectangles. For the first time I was quietly confident of securing a ribbon and the acclamation of my peers.

On the afternoon before the big day, I scampered home for a final flurry of rectangle walking and grooming, but Barry, who normally came gambolling out to meet me, was nowhere to be seen.

'Barry,' I yelled. 'Baz!'

Nothing. There was only a yawning emptiness in his pen.

Then, from the corner of my eye, I saw my mother coming from the house with a strange expression on her face.

'Andrew,' she said to me, because that's my name, 'I'm afraid I have some rather bad news for you. You see, little Barry was very sick, and I'm afraid that this afternoon he passed away.'

'Oh,' I said. 'That's inconvenient.'

Never again would I lead another lamb.

Eventually, after a few years of suffering the humiliation of having to enter the handicrafts competition with my sand saucers and aqua jars, I did re-enter the show ring, but this time with a baby goat. I amazed myself at managing to secure a ribbon for placing third, although there were only two other children who entered goats that year.

It wasn't what could be called an illustrious history in Calf Clubbing, and I thought that the trauma of it was something I had bundled up into a tight ball of guilt and rage, which was wrapped in a thin skin of abandonment issues, and then swaddled in a lifetime of underachievement, and sealed behind a closed door in the back of my mind.

I was mistaken.

A few months prior to the Helensville show, it was decided that for the *Homegrown* episode that celebrated the dairy cow, I would return to my old primary school and finally enter a calf in the Calf Club event.

This seemed like a good idea to everyone involved, except me.

I could only begin to imagine how weird it was going to be to parade around the ring as a 36-year-old man in competition with school children. It wasn't like I was the prodigal son of the district, returning with my own fatted calf. I didn't even have a calf.

Rather than trouble Mother with grooming one, a local woman, Mrs Irving, who was organising the event, offered to provide me with one she had prepared earlier.

Mrs Irving's son, Bevan, had been a childhood chum. Her daughter, Cindy, was the first girl I ever kissed. Admittedly, 'kiss' may be too evocative a word to describe the quick, chaste peck that it was. Nevertheless, Cindy loudly denied it happened whilst we stood in the line to class afterwards. Or maybe she told people and I denied it, I can't remember. It was too traumatising, especially as she was the last girl I would kiss for several years.

Mrs Irving led me to the calf, a small Jersey that I named Minxie.

In order to get Minxie up to show standard, I had to remove a large quantity of what I can only describe as clumped organic matter from her rear flanks. Mrs Irving laughed and pointed out places I had missed.

So there I was, the glamorous and sophisticated metrosexual

media personality, making a triumphant return to his old primary school, camera crew in tow, washing the dried excrement out of the hair of a calf with his bare hands. This was not really what I had in mind as a school reunion.

Thank you, Mrs Irving.

At this point, Mrs Irving, not unattractive herself, decided to show me a little technique known to the inner sanctum of the Calf Club world as 'the lady tickle'.

There is no truly polite way of describing the lady tickle, so I shan't bother with the niceties. The lady tickle involves stimulating an area directly below the calf's vagina, causing her to urinate spontaneously, thus ensuring that she would not do so over me as I groomed her, or in the ring while being judged.

I know! I was shocked too. There are so many levels of abject wrongness in this that I don't quite know where to begin charting them.

I wondered why I hadn't known this sooner. It was a technique I would go on to use many times on my cows in order to stop them polluting the milk bucket. It never ceased to amaze camera crews with its inappropriateness.

With the tickling mastered and the grooming completed, I tried a little preliminary leading. Minxie seemed quite content to saunter along beside me.

What could possibly go wrong?

Well, calves are calves, and calves will eventually grow into cows, and we all know what cows are like.

It all began well. Minxie strutted briskly forward, but at the corner took unkindly to the odd 270-degree turn she was asked to execute, and decided that she did not want to walk any further. I didn't want to haul her forward, as dragging a calf bodily around the ring by its halter is never a great look.

As a child I had found that when I was required to get a recalcitrant calf to where it needed to go, pushing it backwards did the trick. This wasn't an option in the ring. Calf Club judges, like dictators, like their subject to be easily led.

A battle of wills ensued, and we finally completed the circuit and exited the ring to a smattering of polite applause, thankful the worst of it was over.

Only it wasn't.

Derek, the director of *Homegrown*, then decided we needed more shots, and I would have to do it all over again.

The only person looking forward to this less than me was Minxie.

She once again stopped, hauled herself away, and became bovinely belligerent. Thankfully, Mrs Irving sprang to the rescue again, surreptitiously prodding the calf firmly in the rump with her judging pen, prompting it to move forward.

Again, a smattering of applause.

Could this get any worse?

Yes. It could.

I lined up with Minxie and the other entrants in the centre of the ring for inspection. Compared to the other calves, Minxie was a straggly mongrel. The judges in their white coats, armed with their clipboards and furrowed brows, analysed and inspected us. I hadn't been scrutinised this intimately since boarding school, when we had to parade next to our beds before breakfast to make sure we were correctly attired and our areas were clean. I hoped Minxie's areas were still clean.

After much consultation, the results were announced.

I did not win, which came as no surprise to me.

What did surprise me was that I was disqualified on a technicality.

Quite why no one had mentioned that taking your hand off the halter was cause for instant elimination I will never know. Usually I am asked to take my hands off things, not leave them on.

I hadn't realised there were so many rules.

I did, however, experience the belated but exquisite delight of being awarded a nice shiny satin ribbon for participation.

Those who bemoan the awarding of prizes merely for taking part can say what they like. I think it's a good idea.

Back at the Helensville show, the rain has arrived with a vengeance and the people haven't. It is, quite literally, a washout.

Umbrellas are being turned inside out. Gazebos are being blown down. Tempers are fraying in the huge queues for coffee. People are trudging around in raincoats and gumboots. It's too wet for the sideshow rides. Young women in jodhpurs take their lives in their hands riding their horses in a perilous equestrian rink. There's a run on bacon butties. Organisers are walking around looking increasing glum.

The only place that is packed to the gunwales is the indoor produce and craft display areas. People stare at the exhibits as they trudge lethargically past them. The room is hot and a little cloying, as the heat of people's bodies dries the moisture off their clothes. It is steamy, almost subtropical, with an aroma of wet wool and hotdogs.

I am pleased that people have been driven indoors as it will give them a chance to marvel at my success. A plethora of certificates are placed next to my entries, proclaiming me to be a grower of award-winning produce.

Admittedly, the second I receive in the parsley section is tempered by the fact that there are only two entries in that category, but my watermelon and lettuce both come second from a pool of three, and my eggs come third from 12 entrants. The judges comment on the superior yellowness of their yolks.

Silverbeet does, it seems, have a use, as the more of it the chicken eats, the yellower the yolk of their eggs.

I also receive an honourable mention for my corn, which the judge has assumed is some kind of entirely new species. It isn't. It has just been afflicted by me.

As I had suspected, there is one old chap who takes out a prize in nearly every category. In some categories he even receives more than one placing. The diligence and dedication of a lifetime of growing is to be seen in the cluster of certificates that bear his name: Sam Sansen.

I can't imagine how many hours he has put into getting his produce to its prize-winning state. From butter beans to zucchinis, he is the champion. I am shocked to see that his name is even plastered on prizes in the best blooms section. Is there no end to his talent? Flowers as well as vegetables? The audacity of the man.

As I haul the caravan back out onto the highway, I know that I will sleep easy tonight. I'm an award-winning grower of vegetables. Those certificates will be going straight into my CV, right next to my Calf Club participation ribbon.

26

Two little pigs

Thankfully, I am not the only one who is caught eating out of the pig bin. While it would be remiss of me to mention that the other person was Frank the soundman, I wouldn't like you to labour under the misapprehension that we were caught with our snouts in the pig trough as such.

It's just that Mel the producer will often pick up for the pigs day-old bread from a rather nice bakery, which seems to produce an abundance of French sticks. These are not, even at a day old, remotely stale. Can we be blamed for eating them? Everyone loves a fresh French stick. And how much more sustainable can that be? It isn't as if the pigs are being hard done by.

If being caught eating their food degrades my dignity, it is nothing compared to how the pigs ransack their own. I have two on the farm, called Willy and JT, and neither of them has a scrap of dignity. None. Not a skerrick.

There are people who love their pigs, but really, as is often quoted, the pig is a dirty animal.

This fact is reinforced on a warm summer afternoon when I decide it is time to give the chicken house another of its irregular cleans, as a roost full of chickens does produce rather a lot of manure.

A large quantity of this has fallen down behind the netting at the rear of the coop, onto the piece of four-by-two that runs along the bottom of the wall, about 15 centimetres above the ground. The resulting gap leaves enough room for a little ventilation, and provides a convenient place to scrape the fowls' foulings onto the ground outside so that I can then collect it and place it in the compost.

Even as I begin tentatively dabbing at the pile with the spade, I suspect I might be making a mistake, as this wall backs onto the pigpen. Sure enough, within seconds, as the chicken manure falls behind the wire and into the pigpen, two inquisitive snouts appear, sniffing and snorting, and then snuffling the tailings as if they are a prized delicacy.

It is a sight of enduring horror. From that point on, I know that no matter how good Willy and JT will eventually taste, the sight of their unbridled enthusiasm for chicken stool will for ever taint my enjoyment of their meaty little bodies.

It's little wonder that some cultures won't eat pig.

It's also little wonder that some people get it into their noggins that you can feed animals manure. I once saw a study produced by the Food and Agricultural Organisation of the United Nations which was introduced with the words: 'This study describes the potential of nutrient recovery from animal wastes in integrated animal feeding systems.' The study declared that animal wastes represented a vast reservoir of cheap nutrients, particularly for ruminants such as cattle. They even had a name for it — wastelage.

Now, while this no doubt makes sense to the scientist, it just seems to be a little beyond the pale for the ordinary person. Of course, it's nothing compared to the feeding of leftover cow meat to cows, which is one of those practices that causes one to ask: 'Who on earth ever thought that was a good idea?'

Quite a lot of people apparently. It's a real meat-go-round, as cattle are fed to pigs and chickens, which are then fed back to cattle.

Here may be a good time to point out what I believe is the single most important change in thinking that we need to have when it comes to food. Until now, the prevalent mantra has been 'we are what we eat'.

This slogan is something of a deception. The prevailing theory that we need to understand in our modern consumer-oriented world is that 'we are what we eat eats'. And we always have been.

I realise that from some scientific perspectives, the process of turning excrement into stock feed can be considered acceptably safe. I also realise that from some scientific perspectives, the process of feeding cattle to cattle can be considered acceptably safe.

It's just that the concepts themselves, from a layman's point of view, seem unpalatable, and indeed unacceptable.

I have enough people telling me I'm full of wastelage without being fed animals full of it as well.

Except that now my animals are full of it. Or at least the pigs are.

Bother.

I first encountered Willy and JT at the pinnacle of their cuteness. They were eight weeks old, pink and frisky, in a pen in Shawn's woolshed with around a dozen of their siblings and their mother. She was a behemoth. Standing thigh-high to me, and longer than I am tall, she was a solid, low-slung mass of meat and teeth. And quite possibly about the most fearsome animal I have ever encountered.

As she seemed to take exception to me poking and prodding her offspring, I thought it prudent to retreat from the pen and point at the pigs from behind its reassuringly solid walls.

Pigs are an incredibly dense animal. Even as piglets, there is little give in them should you poke one, and poking the piglets was about the only thing I could think of doing to try to assess which two I wanted to take home to fatten up.

'You want one with a nice long back, like that one there,' said Shawn, pointing at a piglet that bore an uncanny resemblance to all the other piglets, and which quickly disappeared again into the anonymity of the squirming mass.

I bowed to his judgement, and Willy and JT were mine.

I have sworn to myself that I will attempt at all costs to avoid what I have long considered to be the most embarrassing cliché of the rural television programme: the hapless farmer chasing the fugitive pig. I now know why these scenes are in programmes.

While they have wonderful personalities and, apparently, the intelligence of a dog, the two things that seem to occupy the vast majority of the pig's time are eating and escaping, no doubt to a place where it expects to find more to eat.

For much of the time, I feel like the commandant of Colditz, trying to stay one step ahead of a diligently determined foe whose sole goal is to loose the bonds of captivity and run free.

The first time they escape is the first time they are ever out of a cage, and for a short while it is delightful to witness.

I love watching young animals that have been reared indoors experience their first taste of freedom. Calves are particularly hilarious. On my parents' farm, calves were raised in barns for the first few weeks of their lives. When we released them they would swiftly realise there was nothing stopping them, and off they would run. They had what could only be described as unbridled glee at the sensation of finally being able to cover some distance. There was a sense of palpable wonder in their discovery of their new environment. They would run and run and run until they ran into a fence. They would then shake themselves off, and run some more.

So it is with Willy and JT. Around and around they run in the orchard. I wouldn't have minded, but I have yet to complete the escape-proofing of the orchard, and if they break out from there, the world is their oyster.

And so I find myself being filmed chasing pigs.

There can be no daintiness involved in catching them. They are deceptively fast and exceedingly elusive. They also have a wonderful defence mechanism: they scream a scream of primal guilt-inducing terror as soon as you grab them. One would think that you were killing them.

After much frustration, I finally manage to grab Willy by the back leg and drag him bodily back to the pen. Swinging him inside, I shut the door and scurry after JT, who is standing panting at the other end of the orchard.

As I am closing in on him, there is a pitter-patter of little feet as Willy dashes back past me. Somehow he is out again. It is at this point that I discover how they have escaped. They have chewed a hole through the concrete wall. I hadn't expected that level of commitment. They are cleverly satisfying both of their desires simultaneously.

I leap at Willy as he passes, then, as JT thinks that I am distracted by his sibling and attempts to flee, I grab him too. Now I have a hold of the legs of two strapping piglets who have no intention of remaining held if they can possibly help it. This is not a vast improvement on the situation.

They both decide that running in opposite directions might help, before turning and running the other way, which causes me to become entangled in my own legs.

I am a knot of man and pig. There are yogis the world over who practise for years to entangle themselves as thoroughly as I find myself entangled. And all the while the pigs are squealing their ear-splitting squeal.

Determined not to let them go, I somehow manage to drag them back to the pen and lift them unceremoniously inside. As soon as I release them, they stop squealing and stare at me with a look in their eyes that says, 'What did we do?' before snuffling around the edges of their pen, looking for something to eat.

Over the course of the next few months, they will escape innumerable times, resulting in the same madcap chase to get them back. It is a few days before they are due to be harvested that they escape for a final time. This time they manage to upset a bucket of paint that is being used to paint The Chapel, which they then proceed to roll in before trotting off to flomp down in the sun.

Jane hasn't noticed the upset paint, but her toddler has, and if it is good enough for the pigs to wallow in, it is good enough for him.

If there is one thing that can be learned from this, it is that paint is quite difficult to remove from a child whose screams rival even those of the piglet.

Pigs seem to be able to consume with ease as much food as you can source for them. Thankfully I have a lot of visitors who regularly come out to the farm. Producers, crew members, researchers, partners of friends of people involved — all are only too willing to bring out their unwanted scraps. I have also approached a couple of cafés in Helensville, and so the routine of collecting the pig buckets begins.

It is something of a demoralising experience.

This demoralisation doesn't come, as you might expect, from having to repeatedly turn up at the cafés to cart away their buckets of waste, while casting longing glances at the people eating their eggs benedict, as the heady aroma of fresh coffee fills the air. Instead, it comes when the buckets are emptied and I have the opportunity to stare at the delicacies that the pigs are eating. They get to dine on a lot of the things that I don't: burger scraps, falafels, bread, paninis, all manner of café leftovers. Sometimes there are even large pieces of cake in the bins.

There is also a certain irony in the fact that the pigs take a lot of looking after and eat a great deal, at a time when the price of pork is at an all-time low. It has fallen below the cost of chicken to become the cheapest meat available as foreign pork is being dumped on the New Zealand market.

It isn't because of this that Mel the producer has been adamant that we shouldn't get a pig. Rather, it is due to the fact that she has a soft spot for them. Most people think they are cute, and express anxiety that these endearing little guys are destined to be killed. Whenever that happens, I politely ask if they want to feed them. People usually leap at the chance and toddle off with a bucket and an air of benevolence.

Invariably they return, ashen faced, asking, 'How soon before they go?' It seems there is little as off-putting about a pig as encountering the voracity of their appetites.

When entering the pen, the first thing that Willy and JT do is to come trotting up on their little hams to sniff and nibble me to see if I am indeed a source of food. Were I to fall over, I no doubt would be.

Often when I am in the pen, fixing the fence so that they can't escape, I find myself squatting down, and the cleft in my buttocks, exposed as it is above the top of my shorts, is given a thorough nuzzling by their damp little snouts.

It isn't an entirely unpleasant sensation; it feels oddly ticklish and amusingly inappropriate.

In the end though, they do make for some mighty tasty eating.

27

A bull at the gate

I will never forget how surprised I was at the rate at which a cow's uterus contracts after calf birth. I witnessed this biological phenomenon as I stood clutching the slick organ in my hands, while the vet, who was herself heavy with child at the time, hastily sewed it back up after a successful bovine caesarean section.

It was a quite an experience for a nine-year-old.

My fascination with the procedure was rivalled only by that of my dog, Spot, who stood looking on, pensive and pleading, timidly edging forward for a chance to snatch a potential titbit. He seemed perplexed as to why we were putting everything back in again.

One has to become quite intimate with animals as a farmer. Take cows for example. If you're not scratching their heads, or massaging their udders, you're thrusting your hand up their jacksies. There is probably a scientific term for this procedure, but I don't know it.

I think of this as I stand in the middle of the paddock rubbing Vaseline onto Harriet's cracked nipples in order to ease the discomfort from her split teats. Judging by the enormity of her udder, she is due to calve, and the warm summer weather is causing the splitting.

I can sympathise. The skin on my fingers is splitting too. It is quite painful.

I always admired my father's hands. They were firmly solid, and roughened by years of farming. My hands, formerly the white, soft hands of a city dweller, have now morphed back into the quasi-rugged land-hands I had been used to as a boy.

What I wasn't expecting was that they would cause me such aggravation. Blisters I expected. Calluses are unavoidable. But having the skin splitting is not something I was prepared for. The small splits prove annoyingly painful — like a multitude of paper cuts. The ones that afflict the top of my thumb and index fingers are the worst. Every time I want to grasp something, or put my hands in my pockets to retrieve something, the splits get knocked and reopen.

My skin is a network of crisscrossing lacerations. I don't know why I expect Vaseline to fix Harriet, when I am applying it with my fingers and they get no better.

As it is difficult to induce a cow to give birth on cue, I am alone at the farm when Harriet goes into labour. There's not a lot that can be done to help at these early stages. All I can do is check on her occasionally, casting a mildly experienced eye over things and noting the discharge, the dilation, and the slightly agitated look in her eyes. Goodness only knows what she is thinking.

What I'm waiting for is the appearance of feet. This is the first real indication as to whether things are going smoothly. They should be the front feet, with the hooves facing down. Shortly after this, the calf's muzzle and head should appear, then its shoulders, and then it should fall out.

If the back feet appear, then it is a breech birth and the chances are it could well be problematic.

As a child I attended large numbers of these births and there is nothing remotely gentle about hauling the calf out. It's brutal. I am hoping, for everyone's sake, especially Harriet's, that things will go smoothly.

Having never had children of my own, I haven't experienced the human variation of the process. For some reason none of my friends have seen fit to invite me to share this joyous experience with them. But I can imagine that it's not all candles and huffing together and wiping tiny beads of sweat off your beloved's forehead, as the two of you hum madrigals and think of baby names.

As night cloaks the proceedings and there is no progress, I retire to the caravan. Harriet seems to be becoming a little sick of my voyeuristic interest in her nether regions. I check her before sleeping, and again when I wake at 3.00 a.m.

I feel as if I should dawdle around the paddock in the dark and see how she is, even though there will be little if anything I can do at that stage. She looks at me with a look in her eye that seems to say that this whole thing would be a lot more appropriate if only I were wearing pants. But it's summer, and the nights are muggy.

At dawn she is standing next to a bright-eyed, little ginger-haired calf.

I'm a father!

The tiny calf has obviously been in the world outside her womb for some time, as he is standing, and clean of any placental residue.

Mother and son seem to be doing well.

He's looking for her teat, but her gargantuan udder is slung low under her belly and he is searching too high. She stands patiently until finally he latches on and devours the warm colostrum.

Milk is dripping from her teats, as she is producing far more than her new boy can ever hope to consume, and I am soon required to milk her to relieve the pressure. She allows me to do this as she stands in the middle of the paddock. What a remarkable cow.

As I strip one udder, the calf is guzzling away on the other side, tail flapping, mouth firmly clamped to a teat, a creamy froth on his lips.

Even as a very young child, I was no stranger to where calves came from. They came out of cows. And how were they put in there? Well, the artificial breeding technician placed them in there by inserting his/her arm into the cow, then injecting the seed that would produce the calf from a long, thin, clear straw they had just removed from where it was clenched in their teeth.

The straw had come from a large stainless steel barrel the technician had arrived with. The barrel, encased in a wickerwork basket, contained lots of straws of calf seed in a liquid that would freeze your finger off in an instant if you were foolish enough to put it into the container.

Oddly for the danger it posed, you never really heard much about liquid nitrogen accidents, nor any crimes being committed with it as a weapon.

'I've got liquid nitrogen — freeze!'

Witnessing the miracle of cow impregnation happening on so many occasions has, I suspect, rather demystified the notion of the reproductive act as a tender moment of intimacy.

Only recently did it occur to me that I had never even thought about the process that led to the capture, in a straw, of the seed that would be planted.

I comprehend this as I stand watching a man masturbating a bull.

Actually, it appears to be a job that requires two men — one to perform the physical act of releasing and capturing the ejaculate, and another man to hold the head of the bull as it rides a dry mount and to make noises of a bovinely erotic nature while it does so. I am not sure whether making the noise is part of his job description, but the man seems rather good at it.

These chaps aren't amateurs. This is what they do for a living. They are just two of the friendly folk that work at the farm and seed extraction facility of LIC (Livestock Improvement Corporation) on the rural outskirts of Hamilton.

Inside its tree-lined acres live the most virile and perfect examples of stud bulls in the dairy industry, whose seed will fertilise cows' wombs for years to come.

It's not until I witness the procedure that I realise I may, if it is indeed possible given what I have just described, have romanticised it slightly.

The dry mount is a small bull or steer, used for the purpose of giving the bull something to become excited by and mount. Their tail is tied down so as to avoid any accidental entry.

The bull mounts the steer several times, until such time as he is considered suitably aroused. Then it's time to introduce the collection receptacle, known as the AV, or artificial vagina.

The AV is about 50 centimetres long, and tubular shaped. When I first saw it, a large industrial thermometer was violating its open end, as it's crucial that it has the correct internal temperature so as not to shock the bull into failing to complete his task.

And they are valuable bulls. One bull, known as 96329 SRB Collins Royal Hugo, or more simply as 'The King', has been responsible for over 1.5 million inseminations over the course of his working life, and is estimated to have earned the dairy industry over $20 million.

This all happens with not a cow in sight. The last cow these sires ever see is most likely to be their mother as they are taken from her and shipped to the facility. Here they remain, living in gay abandon, until such time as they are deemed by a ruthless statistical analysis to be surplus to requirements and are transformed into burgers.

As for the men who collect the seed? I have no idea what inspires them to apply for this most necessary of jobs. Does it begin as a hobby? Do they confess to doing the job they love? What part of their work can they take home? What do they put down when asked to state their occupation? Bull handler doesn't seem quite right.

Don't get me wrong, I am not saying performing the task is shameful. Indeed, had they asked me if I wished to have a go, I most certainly would not have turned down the offer, if for no other reason than that I could say that I have. And who wouldn't want to be able to say that?

This kind of lifestyle is not to be an option for Harriet's healthy baby boy. I have decided to call him Barry, after auburn-haired gardening goddess Maggie Barry.

He will remain healthy, but there will come a time when he will no longer be a boy. Of all the animal interaction to be had on a farm, castrating calves is one of my least favourite activities, possibly because of some subconscious resistance to the removal of such a symbol of masculinity.

I could never participate in cutting them off, which seems too barbaric, but it is certainly quicker than the slow withering induced by the rubber ring method, which cuts off the blood supply until the testes gradually atrophy and fall off.

I've known people in relationships like that.

I have to confess to being quite surprised at the warmth of Barry's scrotum as I cup it in my hand and test its weight. It is firm and slightly furry. My surprise manifests itself in the spontaneous uttering of the phrase, 'Oh, it's warm', which causes great mirth amongst the film crew, and general mocking of me.

While in Mali I had the great fortune of being taken to the oddest religious site I have ever encountered. High on an escarpment above a village, an overhanging ledge protects an area of flat dirt. The cliff walls that surround it appear to have been extensively graffiti'd, but rather than the illicit scrawlings of a dispossessed youth, it is instead the artwork of generations of the young boys of the Dogon tribe. In the middle of the area was a stone, worn smooth by the rubbing of thousands of squirming naked buttocks, as generations of boys wriggled and winced their way through the coming-of-age ritual of circumcision.

It made me glad that in New Zealand the most we might do on our 18th birthday is to skull a yard glass of beer.

For some reason, the job in Mali is done by the blacksmith, which lends itself to a wealth of questions. Given that this is not his primary area of expertise, where would you want to go in the general scheme of things? At the front of the line, to get it over and done with? Or near the end when he is more proficient? Would he become merely workmanlike in the middle, and become casually offhand? Would he sharpen the knife between encounters? And more importantly, what happens to the remains? It seemed impolite to ask.

The initiates have to stay on the ledge for a couple of weeks post procedure, hence the art on the rocks. There's little else to do except heal and paint.

The ledge looks down over the village, and my guides pointed to the house where the female circumcision was carried out. Like the practice itself, it is shielded from people's eyes by being conducted indoors. I never asked quite how much female flesh was excised in that ceremony. I could only hope not much. The girls apparently didn't then have a month off to do arts and crafts; they went straight back to work.

There's little in the way of ritual when it comes to ministering to Barry. The ring stretches out just far enough to almost — but not quite — allow me to slide it easily over his testicles. A little judicious massaging is required, and before the ring is released you need to ascertain that both of the plums (as they are known in the country) are located below the ring, in what Shawn charmingly refers to as the purse. Then the ring is released, it squeezes the top of the scrotum, Barry twitches, I twitch, and all is done.

And why is it done? Well, steers — as castrated bulls are known — are a lot easier to handle than bulls in their complete state. There is no rampant male aggression, no fighting, or breaking fences, digging holes, or accosting hapless cows and impregnating them when they shouldn't. They also grow differently — taller and with more elegant, feminine lines.

However, for all of those women who over the years have accused men of having their brain in their pants, well, in my experience, the stupidity of the steer may just prove them right. I have always found them to be a little bit dumber than they should be. It's most disconcerting.

As a child, I remember once opening up the scrotum of a dead bull to reveal his testicles. They were about 15 centimetres long, glistening white, with smudgy black veins in them. They looked like giant, shiny jellybeans. I left them on the ground, turned my back to do something else, and when I looked back I saw the last of them disappearing whole down the throat of my dog, Spot.

How he managed to swallow them both I shall never know, but if a fox terrier is feisty at the best of times, imagine what he's like with two bull testicles inside him.

I don't think he slept for three days.

28

Playing possum

If you have ever smelt a possum, then you will have a fair idea of what they taste like. They taste not dissimilar to the way they smell. Pungently musky. More disconcertingly, their carcasses, stripped of their skins, resemble those of a little man, or a monkey. It's like eating a leprechaun.

I have been hunting possum as a food source since the day I moved to the farm, but with little success. Rather than looking to shoot them, or employing the simple drive-around-at-night-and-run-them-over technique, I am attempting to catch them with the Timms trap.

The Timms trap is a bright-yellow box equipped with a spring-loaded metal bar that kills the possum instantly by breaking its neck or crushing its windpipe. Either way, for the possum, the result must come as a disappointing surprise, but it is considered to be one of the most humane ways to dispose of them.

The best thing about these traps is that when you go to unload the trap, the possum is dead, unlike a cage trap. My mother employed the baited cage as a way to capture possums. The problem was that you then ended up with an angry possum in a cage. What on earth do you do with them then?

When my mother caught possums, she would either have them shot or she would drop the cage containing the possum into the water trough.

I used to think this was a little rough, but while recording my

Dispatches from the Provinces series for National Radio, I decided to ring the Department of Conservation and ask them what they recommended. I informed them that I had a possum in a cage and that I didn't know what to do with it. I could barely believe my ears when, without a moment's hesitation, a silken-voiced woman responded by saying that I should drop it in a creek. She added that once I was positive it was dead, I should ensure that the cage and the possum were removed from the water, as I wouldn't want to pollute the creek.

Crikey.

As gruesome as that sounds, the Environment Bay of Plenty website says that should you capture a live possum, you should drag it out of the cage by the tail, making sure to leave its front feet on the ground so that it will not be inclined to clamber up the nearest incline, which will most likely be you.

This will also distract it long enough for you to deliver a blow slightly forward of its ears, with a big stick or a hammer, stunning the critter long enough for you to lay it on a firm, hard place and deliver, with your stick or hammer, the coup de grâce. This is not a technique for the faint-hearted.

It is, however, less brutal than their recommended method of disposing of magpies, which is by smashing their heads into a post.

They are a different kind of folk down there in the sunny Bay of Plenty.

Initially, I wanted a possum for its meat. Or rather, Jane wanted it as meat for me. She thought it would have a suitably sustainable aura.

We hadn't counted on its unsustainable aroma.

Now though, I want them for their warmth.

A friend of mine bought a Timms trap to catch a possum that was plaguing a tree at the bottom of his section.

He caught the possum and reset the trap on the off-chance there was another one. There was. He continued to set the trap and catch them, until eventually he decided he should do something with the skins. He tanned them and then had them fashioned into a blanket. It was luxurious, seeming to generate its own heat when you lay on it.

I'm not planning anything as extensive as a blanket. All I want is enough skins to fashion a hot-water bottle.

The reason I want a possum-fur hot-water-bottle cover is that the caravan is freezing and I can hardly install the potbelly in there, although the thought has crossed my mind. Nor have I enough electricity to run a heater.

I do, however, have enough solar electricity to power my laptop and some LED lights. The solar system has been installed by Walter, a sun lover from Germany. By sun lover I mean, of course, a solar enthusiast, not a nudist.

At least, as far as I know he isn't a nudist. He might be, I guess, but that is really up to him.

He was certainly wearing pants when I visited him at his house one overcast day to discuss installing a solar set-up that would produce enough current to power my laptop and recharge my cellphone, because even when living in a field there are moments when those two appliances are important.

We chatted about the idea over a cup of tea made from water that he had boiled with the power of the sun, harnessed by tin foil, which when laid in the correct concave shape has the ability to focus the sun into one heat-inducing spot.

He thought that my solar power ambitions would be no problem at all, and within a short time he had installed a system.

Now, for the first time in months, I have the ability to generate light other than with a temperamental battery or gas-powered lamp. Happy days are here.

However, the LED lights emit a cold blue light that makes you feel chilly all the time. I would have preferred them to have been warmer. Red would have been awesome, as it would have made it seem as if I were on the bridge of a submarine permanently at action stations. I would have had an enduring sensation that something could happen at any minute.

I tolerate the LEDs for some time until one night I am struck by the thought that as I am able to charge my laptop, there is no reason why I can't plug in a lamp. I haven't told anyone else about the length of time it took to come to this revelation, because it might make me appear a little dim.

I did wonder if the system had the power to electrocute me. That would have been an embarrassing way to go.

I have also decided to reduce my battery use when powering little essentials like my transistor by investing in a wind-up radio. It is excellent. No longer do I have to suffer the inconvenience of flat batteries. I simply wind it up.

Instead, I now suffer the inconvenience of having a tranny that isn't very good at picking up stations. Every station sounds like it is being broadcast from the moon, which is also, inexplicably, under water. I have to hold the tranny continually, usually in a position of mild discomfort, in order to hear the news headlines.

But the solar power system is unable to support a heater, which means I am still cold.

At one stage, it occurred to me that I could be warm in a garment knitted from the wool of my sheep.

I'd taken the fleece up the hill to Sharon, a wool carder. Fittingly, she was a card. Hailing from Australia, she was a bundle of energy. One only had to ask something and she would do it. 'No problem,' she would enthuse.

She was a necessary cog in a home spinning wheel, because in a small shed on the back of her section Sharon had installed a wool-carding machine. She would wash the wool, and then the machine would brush the wool ready for spinning.

She invited me to join the spinners.

In a small church hall, as the twilight became night, I sat with the gentle rhythm of the spinning wheel, clickity-clacking, as the toes of several women of a certain age rhythmically massaged the treadles of their spinning machines. They were kindly attempting to teach me to spin, a task that seemed to involve a little too much multitasking for me to fully master. I would master one part of it, but would then forget to breathe. When I remembered to breathe, the part I had mastered fell apart.

I had baked them an apple crumble in the mud oven, and I was hoping that I hadn't dropped any of the rock-hard door grit into it, as it had the ability, fired concrete-like by the heat, to break a tooth if it was bitten. That would have put a damper on the evening.

By the end of the evening, I had a small ball of twine of no specified nor uniform thickness, and a bag of carded wool that appeared not to have been touched.

This was disappointing as I had wanted to have spun enough wool to have been able to knit some glittens — half-gloves, half-fingerless mittens. The glove portion folds over the fingerless fingers to keep them warm. The small flaw in that plan was that at that stage I wasn't quite sure how to knit.

I failed, too, in converting the rest of the wool into felt. I thought perhaps a nice felt poncho could have been a snug winter fashion item. I managed to ruin a large hunk of my carded wool with soap and water before Sharon informed me that it was not the right kind of wool for making felt.

Either one of these sadly non-existent garments would have been perfect for the situation where I now find myself with Shawn, one icy night. As I've had no luck with my Timms trap, he has offered to take me to a forestry block he is overseeing to see if I can shoot some possums.

'There might be few deer here too,' he says. 'We could get you one.'

Sure enough, on the brow of the hill at the edge of the headlights' brightness, the fleeting shape of two young stags sweeps through. They're a majestic animal, whose speed and grace make them appear as if they are flying low over the landscape, vaulting fences, gliding over scrub, and disappearing in the blink of an eye.

'If we do go for a deer, you'll have to shoot it in the head or the neck,' counsels Shawn, 'and you won't get much time. If you can't get a clean shot, don't shoot.'

I couldn't agree more. There's something noble about the deer, and I don't want to wound one. It will be a clean kill or nothing at all.

But it is not to be. By the time I have exited the vehicle, the deer are spooked. I barely have time to line one up in the scope and it is gone, disappearing silently into the night. I don't mind. There is always next time.

There is no such problem with the possums. They are everywhere. Some stand their ground in the middle of the road, others scamper off, some mooch about, whilst others climb the nearest tree as fast as their little legs can propel them.

I want to kill them cleanly too, with a shot to the head, and they seem to know this. They have a superb defence mechanism, which consists of walking slowly away from me with their heads down, so that all that fills the scope is their fluffy rump.

The .22 isn't really powerful enough for shooting right through them, so it is a matter of waiting to see if they turn around to look at what is going on. Sometimes a whistle works, and at other times they march slowly on, turning only when they come to a fence.

I shoot them on posts. I shoot them in paddocks. I shoot them out of beeches. It is carnage. It isn't that I stop counting how many I shoot, but I simply lose count.

Is it simply killing for sport, or do I do it because of some perceived notion that the more possums I destroy, the more I will help the bush? It isn't like I will clear the area of them. A few nights later we will head back out again, and there will be no shortage of them then either. All that we can do is create a vacuum into which more will flow.

So why kill so many? For fun? I really don't know.

One childhood night out hunting, I'd picked up a possum that I had shot by the tail. Spooked by the dark at the best of times, I leapt nearly high enough to clutch the moon when the supposedly dead possum had swung around and scraped my leg. Since then I have been wary of any animal playing possum, pretending to be dead when they are most certainly not.

Especially possums playing possum. They are good at that.

Whilst out with Shawn, I also take the opportunity to bag a few rabbits for a casserole. Shawn checks the rabbits carefully before I shoot them to make sure they are not a little ginger one that he has seen around, and that he and his daughters have informally adopted. Once again, the odd one out is saved by virtue of its freakishness.

I gather up the best possums to take home to skin. I could have plucked the rest of them, as possum fur can be plucked while the corpse is warm quite easily, with the fur coming off in great handfuls.

I've hunted possums for years and never realised this.

I have invested in a bottle of hide tanning solution. This is not a bronzer to make me appear a little more sun kissed, but a product to turn skins into leather.

The instructions are, to be generous, scant. I expect more information, but they seem to assume that whoever is doing this has some kind of knowledge prior to purchasing the formula.

First, I need to salt the skins, which I do by staple-gunning them to a piece of plywood, fur side down, and coating them liberally with salt. Then I leave them to season, mix up the solution, put the skins in, forget to agitate them, take the skins out, and wash and dry them.

It's not a dismal failure, but nor is it the result I was expecting. Clearly it's not as simple as the lack of directions would lead one to believe.

Jane is still insisting that I savour the meat of one, and I have to confess to a certain curiosity. I would hate to think that it was delicious and that I had been missing out on a free meat treat all of these years. At one stage some years prior, my local Chinese greengrocer had been selling them in frozen and bagged form, only they were calling them 'apple-eating tree bears'.

In hindsight, I'm amazed I could resist.

I've always enjoyed eating an animal small enough to be served whole. There's no mistaking what it is. This is the case with possum.

I've checked its liver for any signs of lesions that might indicate tuberculosis, as I have no wish to contract TB and become a frail male Katherine Mansfield, coughing blood as I linger over my work of great New Zealand literature.

There is no disguising the smell though. It's possum through and through.

It's a smell that takes some getting out of the pores of the skin. Having spent the better part of a day skinning a pile of them, I find the scent has worked its way into my skin, and I smell for days. I'm surprised I'm not ravaged in my sleep by randy possums mistaking me for one of their own.

I have looked on the Internet for recipes, and many mention the need to remove the musk glands, located in the small of the back and on its front, but I can't find them. All I know is that when selecting a possum to eat, you have to ensure it's a female. There's nothing like the taint of testosterone to ruin meat.

In the end I merely slather the carcass in olive oil and garlic, stuff the belly cavity with par-boiled potatoes and carrots, and fold it up to fit in the roasting tray. The unfortunate thing is that the corpse lies on its back, with its back legs spread akimbo in what can only be construed as a posthumous tribute to *Hustler* magazine. It is positively exploitative.

Having left its furry paws on like little hairy mittens, I am bemused to note that the hair hasn't burnt off them when I withdraw it from the kiln. What I hadn't expected was that the tail would curl up completely, like a little crispy koru. So grotesquely cute.

There is a surprising amount of meat on the possum, and most of it serves only to prove why it is that not many people choose to eat it.

Not even a considerable amount of the applecohol I have brewed with Rudolph can remove the lingering taste.

The only drawback to the possum-fur hot-water-bottle cover is forgetting you have filled it and put it in your bed. It can seem, as you slide your toes down under the blankets, that something has come in from the wild and snuggled into your bed.

Believe me, that can be frightening, especially when you smell as I do.

29

Lambs to the slaughter

Standing in the door of the local hall with the collective eyes of the room running from the feathered plumage of my headdress, down past the clinging tightness of my sequined mini-dress, and lingering on my shapely, pantyhose-clad legs and high-heeled feet, I realise that I may not be dressed entirely appropriately in my quest to borrow some local children.

Instead it appears that I am the horror of the local menfolk and the envy of the local women, as I have always had wonderfully shaped ladylike legs. It's both a blessing and a curse.

I had volunteered my services as the MC at a Las Vegas-themed fundraiser for the local playcentre, and I'd simply assumed that such an event would entail people dressing to the theme. I was obviously wrong.

As part of the proceedings involve a charity auction, I thought what better way to raise money for them than by auctioning off a chance for some of the children to live the television dream by appearing in my programme for a few short moments. It will raise some much-needed cash for the playcentre, and I will gain a little free labour to help me around the farm. Everybody will win.

Over the course of the next few hours, through a combination of subtle flirtation and threats, I help to empty the pockets of the assembled masses. By the time the auction rolls around I have redeemed myself to such an extent that a flurry of bidding results in the splitting of the prize between two families, who promise to bring their offspring to the farm the next weekend. I spend the following week envisaging myself being carried from task to task on a litter, reclining with a fruity cocktail as I direct the children in their labours. Sadly, when they arrive, they are slightly smaller than I had imagined, and I have to scale back my plans as I don't wish to send them back to their parents any more damaged than is necessary.

The main reason I need the help is that I am in the last days of my sustainability experiment, and I require assistance in a final push to finish such tasks as tidying the gardens before replanting them so that the next occupier of the land has something in place when they arrive. After all, one of the tenets of sustainability is leaving the land in better condition than when you found it, and that at least is something I feel I can achieve.

More importantly though, I need help in wrangling my two remaining sheep. In lieu of a dog, what could be better than a few energetic children? My only concern is that they will prove wildly more undisciplined and difficult to control than a dog.

I need to pen the sheep as their presence is required at an event to be known as 'Te Radar's Last Supper'. What better way to say thank you to the community that has so generously helped me to survive than by impaling an animal on a pole and roasting it over hot coals, so that we can then break bread and share meat and drink booze over the course of an evening.

Of course, I don't think it entirely wise to inform the children of the reason they are rounding up the sheep.

The children have a rather misguided notion of how easy the task in front of them will be as we head up to The Tops to muster the hapless sheep. Clearly they have not met my sheep.

As soon as we enter the paddock, the sheep stand up, then as I encourage the kids to 'Get away back', the sheep turn tail and bolt to the top of the paddock, followed closely by the children.

Up to the top they all run, and once there the sheep run all the way back down again, followed by the puffing children, who, once they reach the bottom, can only look on as the sheep elude them and once again run back to the top of the hill, where they stand watching as the children, encouraged by my shouts to 'Go round', stagger back up after them.

Using all the skills I can recall from a childhood spent transfixed by the iconic New Zealand television programme *A Dog's Show*, a riveting half-hour of sheep dog trialling that would hold the nation enthralled of a Sunday evening as men and dogs pitted themselves against the unpredictable sheep, I try to position the children so that they can bring the sheep back with minimal effort.

'Steady . . . steady,' I yell at them. 'Go round, that's it . . . steady, go right . . . sit down, SIT! Good boy.'

Tentatively the kids work the sheep down the hill and into the pen. I close the gate, and we are done.

'Good kids,' I say, patting them on their tousled heads. 'Now, go home.'

With the kids gone, there's nothing to do but wait until Shawn arrives the following day to help me prepare the sheep for the party.

As the sheep loiter in their pen, nibbling at hay, unaware of their impending fate, I realise that in a way, the Agrodome celebrates only half the sheep's contribution to the country, for it only tells the story of their fleece. While sheep have served us handsomely with their wool, they have also laid down their lives for the meat industry.

Thankfully, this part of their legacy is celebrated at the place that I rate as one of the nation's greatest theme parks, Totara Estate. Nowhere else in the world would I imagine has an abattoir been converted into a fun family day out.

Nestled beneath a hill just south of Oamaru, Totara Estate is the site where the sheep were mustered and slaughtered to produce the meat required for the first consignment of frozen lamb to be exported from New Zealand.

A few of the original buildings remain. One is the original killing shed. As you approach the building, you notice some clay spouts jutting from the building at the level of the internal floor. These served the simple purpose of allowing the blood and offal to flow out of the shed and into large troughs that serviced the half-acre pen reputed to contain some of the most ferocious pigs you were ever likely to encounter.

This was no doubt considered at the time to be a state-of-the-art and highly efficient waste disposal method, and when dealing with dead sheep, there is an awful lot of waste to dispose of.

At one point, I happened to be away from the farm filming *Homegrown*, and Jane had booked in the home-kill butcher to dispatch a couple of my sheep. She had forgotten that the home-kill people only take away the dressed carcasses, and so she was left with a wheelbarrow filled to the rim with the voluminous contents of the sheep's innards. These require quite a large hole in which to bury them, and I'm sure that as she laboured away digging it, she must have been thinking that this was not quite the glamour of the television industry that she had run away from Taranaki to find.

During the summer of 1881–82, six butchers worked inside Totara Estate's killing shed. This was before the era of the modern killing chain which, inspired by Henry Ford's production lines, became the industry norm, so these men would individually kill and butcher up to 50 sheep a day. Nowadays, those who work on the meat chains tend to perform specific parts of the task — indeed, sometimes the exact same few, precise movements, all day, every day.

Today, the killing shed's platform where the slaughtering and butchering occurred is quiet except for the charmingly ghoulish animatronic display that tells the fateful story of the progress of the sheep from the estate, through the killing line, to the waiting ships.

The tour of the killing house follows the same route the sheep took, from the kill floor, through to the superbly named Carcass Hanging Shed. This stone-walled building is really a simple room with wooden slats on the floor. Gaps between the slats, combined with ventilation in the walls, allowed cool air to circulate through the room. Even on the stiflingly hot day that I visited, a chilly breeze blew in from under the floor and no doubt up the long skirt of the woman in traditional costume who was guiding me on the tour, as we stared at the fibreglass replicas of sheep carcasses that now hang there.

Hanging next to these gruesomely realistic carcasses are a few of the original tins that contained some of the products that were exported. Quite what anyone would have done with a two-litre can of lamb tongues, or liver, I do not know.

Prior to the arrival of Captain Cook, there were no mammalian animals present in New Zealand other than a species of bat and rather a lot of seals and whales.

When he happened upon the place in 1769, Cook had with him in the cramped confines of the *Endeavour* a few sheep. For some reason he decided not to allow them ashore, and eventually sailed off with them still aboard to declare the country found.

On his second trip, Cook arrived with two merino sheep. The sole survivors of a flock of six, they had endured the epic trip from South America via the Antarctic Ocean. Lovingly nursed through illness, they'd survived the cramped conditions, the threat of the cooking pot, and lecherous sailors.

On 20 May 1773, while in Queen Charlotte Sound, Cook noted in his diary: 'This morning I put ashore a ewe and a ram, for my intention is to leave them in this country.' Three days later, he wrote: 'Last night the ewe and the ram died.'

The poor creatures didn't succumb to hunger or to the predations of hungry Maori, but rather it seems they simply ate some kind of poisonous berries, quite possibly those of the tutu. Cook wrote: 'Thus, all of my fine hopes of stocking this country were blasted in a moment.'

Undeterred, in 1777, on his third voyage here, he released another pair. They too promptly died.

To be fair, it wasn't an overly auspicious start. We should be glad, I suppose, that Cook was the captain of the *Endeavour* and not the Ark.

It was to be another 60 years before sheep were successfully introduced to the country by the Reverend Samuel Marsden. Known as the 'Flogging Parson' due to his reputation for punishing transgressors with a taste of the lash (and not for a penchant for onanism), he performed his first sermon here on Christmas Day 1814. Quoting from Luke 2:10, he said: 'Behold, I bring you good tidings of great joy.' He could have added 'and some sheep, and a few cows', but regrettably he didn't.

By 1858, there were 115,000 people living here, alongside 1.5 million sheep. With little else to do but breed, by 1865 there were over nine million sheep. By 1878 it was estimated that there were 13 million sheep, and it was the considered opinion of people of sound mind that the golden fleece the sheep were producing had an unfortunate by-product — the sheep itself.

Their numbers proved so problematic to farmers in the South Island that it was decided that the only sensible thing to do was to bring herds of them to places such as the high cliffs south of Oamaru, whack them on the head with a hammer, and throw their carcasses into the sea.

Someone once recounted to me a Maori proverb that goes something along the lines of: 'Pakeha treat the sea like a rubbish dump, whereas Maori consider it to be their refrigerator.' While I enjoyed the sentiment of this phrase, I didn't have the heart to be a pedant, and say: 'But Maori didn't have refrigerators . . . '

There was a poignant irony in this proverb, because it was to be refrigeration that would save the day. I might have added that refrigeration would save the bay too, but we found new and better ways to pour yet more effluent into them, particularly from the freezing works that would spring up as the answer to the problematic sheep.

There was no way we could eat all of the lamb that was processed. Fortunately, there was a desperate need for it in the United Kingdom, where meat was scarce, and expensive. All we had to do, as someone whose name now eludes me once said, was to 'subjugate the tyranny of distance' and we would have a licence to print money.

And so it was in 1881 that, for the sheep at least, all roads led to Totara Estate. From there it was but a short rail journey to the wharf in bustling Port Chalmers, where the *Dunedin*, one of the Albion Line's best and fastest sailing ships, sat waiting. Originally designed for the swift transportation of emigrants, it had been stripped out and converted into a new-fangled floating freezer.

Each day more carcasses arrived and were laboriously loaded aboard by hand. When there were about 1000 on board, the crankshaft, which powered the motors that powered the freezers, broke down. So the dockhands laboriously unloaded all the carcasses and sold them to the locals, making the good people of Dunedin the very first recipients of frozen, export-quality New Zealand lamb.

They eventually fixed the motor, and finally on 15 February 1882, the *Dunedin* set sail for England with 4500 frozen sheep carcasses, various cans of tongues, liver and other assorted sheep-based products, and two passengers.

There were supposed to have been more passengers, but a rumour had circulated that the crankshaft might break again, which could well mean the sinking of the ship. This naturally deterred a few people. Of course, if the crankshaft broke and didn't sink the ship, the crew and passengers wouldn't be able to eat 4500 sheep carcasses fast enough to stop a rather unpleasant odour from developing. This naturally deterred more than a few people.

The journey should have been straightforward, but the engine that powered the freezers was situated below deck, with a funnel that came up through the deck, and up through the funnel came clouds of sparks that had the bothersome habit of setting the sails on fire. Such was the extent of the problem that the ship's captain, a man by the name of John Whitson, was worried that the masts would also catch fire.

Also, in order for the freezing units to work, they required a certain degree of air flow, which was fine until the ship became becalmed in the tropics. Captain Whitson did as any good captain ought to in this circumstance, and lowered himself into the refrigerated bowels of the ship in order to bash holes in the vents to allow more air to circulate.

This was not only cramped and tiring work, but it was also rather chilly, and he eventually passed out and began to freeze, resulting in the mate having to lasso his feet and haul him out to be defrosted.

The *Dunedin* finally arrived in England after a voyage lasting 98 days, with only one carcass spoiled. In an age before the concept of food miles was even dreamed of, this was considered a great triumph, and the success of the voyage paved the way for countless billions of sheep to give up their lives for our export returns.

Within 10 years, a fleet of 40 ships, some carrying up to 70,000 carcasses, were plying their trade between here and England, and we had become, as historian James Bellich wrote, 'England's protein factory'.

The story of the *Dunedin* doesn't finish there, however. In the Port Chalmers Museum, there is, amongst its teeming haul of memorabilia, a middling-sized iron key that is reputed to be the key to the door of one of the crew's cabins on the *Dunedin*.

The man whose cabin it was is said to have had some kind of premonition while the ship was in port at Oamaru in 1890. Just prior to it setting sail for England, loaded with sheep, he quietly locked his door, put the key in his pocket, slipped ashore, and never returned.

The *Dunedin* set sail without him, and was never to be seen again. It is thought that somewhere off Cape Horn she struck an iceberg and was lost with all hands, and a lot of meat.

The key is said to be all that remains of the *Dunedin*.

I suppose we can only assume it is a key off the ship, for unless we find the remains of the ship with the door, or at least the lock to the door, intact, then fit the key and see if it turns, it could really be any key.

But it is nice to think it isn't just any key.

Despite the ramifications of all this, and in a country whose economy was built on an orgy of wanton sheep slaughter, the actual process of the killing of a sheep is not something most people usually experience or even think about.

With guests due for dinner in a few days, it is something I have been thinking about experiencing for some time.

Despite all the hunting that I've done during the series, there is an essential difference between the adrenaline-fuelled action of sticking a squealing pig during a hunt, or shooting an animal from a distance with a gun, and opening the sheep's pen as it stands there patiently watching you, dragging it out, and slitting its throat with a knife.

But that is what I need to do if I am to feed my guests. I couldn't in good conscience talk to people about the disconnection between them and the meat they eat if I do not harvest my own.

The sheep have been left in the pen next to the chickens overnight, and when I go to feed the hens, the sheep look at me with large, quizzical brown eyes, as if to say 'Why are we standing here? It's weird.'

I feel a little sorry for them, and mumble, 'Don't worry, you'll be fine.' I think it is an indication that maybe I am worried I won't be.

You can't really practise killing a sheep, and there is a lot that can go wrong and result in a botched job. Shawn has offered to show me the correct way to go about it on the first sheep, and then I will deal with the second.

In the end there is an odd formality and grim sense of purpose about the whole event. Shawn arrives. We set up an ingenious system of poles and pulleys that are attached to the side of his ute to make the dressing of the carcasses easier, and I go into the pen to secure the first sheep.

As I struggle to drag it out, Shawn says: 'You know, if this sheep goes to the works, he gets to go on a long journey in a truck, stressed to the max, gets to the yards, they hose him off, he can smell all this death around him, he stands there while three or four thousand of his mates have to go in front of him, and they slowly work their way through to him. One would imagine that he's not that happy about it by the time they get to him. It's a lot better for him to go like this, happy in the paddock one minute, dead the next.'

It's a good point, and one that has been emphasised innumerable times over the last 10 months as I've worked my way through my livestock.

While some people prefer to shoot the sheep first, others render it unconscious by hitting it on the head before slitting its throat. We are using the more traditional method of simply slitting their throats. This serves the dual purpose of simultaneously killing them and allowing them to bleed out, so that the meat isn't tainted by the blood.

I lay the sheep down and Shawn takes me through the procedure. Within seconds the sheep is dead.

'You see,' says Shawn, 'it's all over before they even know what's going on.'

And now it's my turn.

I negotiate the remaining sheep out of the pen, lay it down and, holding it with my knee, pull its head back by the chin, exposing the throat. Even now it doesn't struggle.

With the knife that Matt gifted me on the day I killed my first pig, I cut firmly across the exposed throat. The sheep's neck is tougher than I expect, and instead of severing the neck in one swift movement, I experience the horror of having to take an extra few seconds to hack the knife through it.

I am completely appalled.

It's not quite the swift, clean kill I was hoping for, but as the warm red blood hoses out, I finally sever the neck all the way to the backbone, and with a final pull of the chin I feel a definite crack as the neck breaks.

The sheep kicks its last and is still. The torrent of blood eases.

It's done.

Slightly shaken, I look up at Shawn and say, 'I think I was more stressed about that than the sheep.'

'I think you were,' he replies.

Turning to Jane, I say, 'We're definitely not going to see that bit on telly, are we?'

'We most certainly are not,' she says.

All that's left is to skin and gut them, and then let them hang overnight from a branch under the totara tree.

If I had really thought about it, I may have found it a quirkily symbolic association with Totara Estate. But it never occurred to me at the time.

With their carcasses swinging gently in the afternoon breeze, I go to light the fire under my bath.

30

Christmas the turkey

I don't want to be known as the man who killed Christmas. The little guy has won the hearts of everyone. Once again, being decidedly odd has saved a life, and they don't get stranger than Christmas the orphan turkey.

There is no denying that he is slightly unhinged. The poor fellow is clearly going through a flush of adolescence, and nothing is safe from his passionate predations. Solar lamps, cans, logs, buckets, bricks, the camera tripod, gumboots — all have succumbed to his amorous advances. The only things that are safe are the chickens, who pay scant regard to his proposals of passion.

I'm sure he cannot figure out what is wrong. His displays are something to behold. From the moment he flutters down from his roost on the corner of the chicken coop, where he sits alone at night, his demeanour indicates that he would like to be loved. His chest feathers puff out, his tail feathers fan out, and his wattle changes from white, to bright red, to blue.

His snood, that long, floppy finger-like protuberance on his beak, elongates and dangles down over the side of his nose. It's quite unsightly. He puffs himself up and lets out various calls, from the familiar gobble-obble-obble-obble to a noise that resembles a boy racer's blow-off valve, a sort of rapid expulsion of air in an abrupt HUMFF.

Men find him repulsive. Women love him.

But no matter how much he displays, or how much he calls, or how much he flirts or prances, there is no one to answer his call.

He needs to move on.

He's not the only one.

For the last few weeks, it's felt decidedly end-of-termy around the farm. There hasn't been any real point in starting any big new projects. No plants planted will mature before I leave. All there is time for is a period of introverted reflection.

Finally though, after months of struggle, I have done it. I am surviving solely on what I have grown.

Admittedly I don't have as much variety in my vegetables as I would like, but I continue to eat exceptionally well. There is more than enough to sustain me through the winter, should I be staying that long. Each night I prowl the gardens, selecting fresh tender morsels.

I did realise at one point that the bitter leaves I couldn't manage to tame weren't a stir-fry vegetable. I was actually eating the wrong part of the plant. How I laughed.

While some people are meticulously diligent with their gardens, I have experimented with the hands-off, low-maintenance, allow-them-to-look-after-themselves types of gardens, and they are stronger for these strategies.

My theory has been that you don't need to be a fanatic to achieve moderate success. The low-effort garden is a good method on which to base a sustainable lifestyle.

I don't want to be seen on national television as a slave to my crops, because it seems to me that there are a lot of people out there who have the same aversion to effort that I do, and television needs to cater to them. Fortunately, they are probably already in front of it.

For too long, we who don't want to do much have been ignored, and I think I have redressed that balance.

To me, the farm has always been about what is possible for the guy who doesn't like doing much, but who has a lot of plans and dreams and hopes and expectations. We can't all be zealots.

It's the sheer amount of work involved in running a farm that's difficult. It is little wonder that monks shut themselves away in closed communities and live self-sufficiently. It's not that they don't want to take a break away, it's just that there simply isn't time.

I say that not as a complaint, but simply as a fact. Statistics indicate that the average length of time that people own their lifestyle blocks is somewhere around 18 months to two years. The lifestyle block really does block your lifestyle.

Still, as well as the vegetables, I'm eating goat and mutton and pork and turkey and chicken and fish and duck that I have produced or hunted. It's immensely satisfying to do so.

I'll be taking a freezer full of meat with me when I leave. That means I'll be running two freezers at home and will be responsible for the associated drain on the country's meagre electricity reserves.

I'm not eating very many eggs though, because I can't figure out if the chickens have stopped laying or whether they are laying somewhere else. At one stage I discovered a clutch of 15 eggs under The Chapel as the broody hen had been stockpiling them. She hatched three. They were a delightful late-summer surprise.

The most important question though is simple: have I become more sustainable? The answer is a definite yes.

But in many ways, the effort of making a programme on sustainable living may have proved to be unsustainable.

In the process of filming me helping the planet, there were routinely at least four and at times up to nine cars at the farm, each of which had travelled the 45-kilometre trip from Auckland.

Brendan the editor, who travelled out each day to edit the day's filming, calculated that over the six months that he spent travelling to the farm, he totalled over 11,000 kilometres. That's the distance from Auckland to Bluff and return three times. Plus some.

When the mountains of camera batteries were put on to charge, I swore I could see the lights of Auckland dim.

When we began this experiment 10 months ago, sustainable living was the hot topic. People were talking about the importance of eating local produce. There was talk of the end of cheap fuel.

Over the period of time that it took to film the show, much changed. Although fuel didn't run out, the price of it continued to rise. Countries around the world began having food riots — pasta riots in Italy, rice riots in Asia, and tortilla riots in Mexico.

With the price of food rising everywhere, people don't care so much about eating local. They just want to eat. Drought is devastating crops. Biofuels, thought to be the saviour of the globe, have in turn been blamed for worsening people's lives as they use up precious farmland and grains that are used to feed stock and people. They are even blamed for the destruction of ever-increasing areas of rainforest, as people scramble to find more land to grow them.

If we do stop using so many shopping bags, will we really make any difference in stopping global warming, if it even exists?

For those who wish to become more sustainable, there are so many dilemmas. It may simply have been too many hours alone in the tent, then the caravan, but there were a lot of problems I never solved.

Take the matches versus the lighter dilemma.

I could have got a refillable lighter, but I keep losing lighters, so why spend too much money on them? The fact that they are made of plastic, contain gas, and are disposable makes them a great example of the wicked modern petrochemical world we inhabit, doesn't it?

Matches, on the other hand, are made from wood. But there must be a lot of forests out there dedicated to solely growing trees for matches. I have no idea what kind of wood they are made from, nor if different varieties are used. What I do know is that if the packet of matches gets wet, as it invariably does, then the matches become useless. So what to do here?

There are many other smaller, but equally mind-boggling issues.

Like what can I use instead of tin foil if I don't have access to banana leaves or suchlike?

What should I burn in my lamp — kerosene, gas, or citronella oil?

Are insect coils carcinogenic? I always imagined incense was, because after all, it does produce a lot of smoke.

I read somewhere that the ground-up powder of roasted beetroot makes for a good coffee substitute, but would that really make for the kind of refreshing beverage I love?

The axes I use have some kind of composite plastic handle. Are these better than a wooden handle? Looked after properly, a good wooden handle can last for years. The catch is to look after it properly. I am not sure I am overly suited to doing that, so maybe a plastic handle is more sustainable for me, providing I don't simply lose the tool somewhere.

I may have children one day, and then I'll face the hoary old cloth-nappies-versus-disposables dichotomy. It's argued that by the time you take in the complete environmental impact of cloth nappies — including electricity for heating washing water and running a dryer, detergents, the effects of monocultural cotton plantations, and the time all this labour takes — then maybe disposables aren't too bad.

Toilet paper is a classic. Could there be a more disposable product than the toilet tissue, the paper towel, and the nose tissue? There's always a hysterical rumpus from the masses when it's suggested that maybe we should use fewer leaves of paper. We're not animals, we cry. But some countries don't have toilet paper and they seem to do okay. We could all go continental and use bidets instead, but then we'd be wasting water.

In reality, who really has the time and the stamina, indeed the sheer bloody-mindedness, to sit down and not only analyse the studies and what they declare the best decision to be, but compare them to each other, research who funded or has vested interests in them, then base them on your situation, and follow them, until another study comes out negating the previous one, or a new product comes onto the market?

There are so many opinions, it's difficult to know who to believe.

The result of this is that we live in an age populated by infinitely more eco-worriers than eco-warriors.

For me, and I suspect for many others, life is too short and too busy (even if we are busy doing very little) for things to be an effort. Sustainability is like a diet, and a diet that is difficult to follow and not very enjoyable won't last. In other words, a sustainable lifestyle needs to be sustainable. And enjoyable.

But it is possible and it can be fun, and there are ways to make a difference without busting a gut. I've certainly seen that. It can be as simple as a few pots with some herbs or tomatoes planted in them, or as effortless as planting a few fruit trees around the section. If you have a pre-mix faucet and you only want to use cold water, turn it all the way over to the cold side, otherwise every time you use it you'll be draining a little of your hot-water cylinder. Tune your car. Buy some chickens.

All it takes is a few small steps. I'll be taking some, which will more than likely be those steps from the car back into the house to get the reusable shopping bags I've forgotten and that the note on the dash reminds me to get each time I hop in the car.

And I'd like, just once, to perfect a compost pile.

As darkness falls on my last night at Te Whenua o Te Radar, a procession of cars travels down the long driveway to come to my last supper. Many of those who have helped and shared and mentored and tolerated me are here.

I get the sense that in many ways, this show has not been about sustainability. It has been about community.

Just as Christmas the turkey needs to be with his own kind, so I have needed the help and support of the local community. Just as he was taken under the wing of a broody hen in his hour of need, I too was taken under the collective wing of a community of people who protected me, taught me and fed me.

There is no self in self-sustainability.

It's time we found Christmas a special lady.

One of my last acts on the farm is to throw some feed in the paddock, wait for Christmas to distract himself with it, and grab him by the legs. It's the first time I have held him since I put him under the chicken. He is deceptively hefty, and after an initial fluttering is content to hang upside down in a state of confusion.

He is going home, not that I suspect he knows it.

Taking him back to where we first met him, I take him out of the cage, and set him free. He stands there and looks at us, as if to say: 'What?'

Eventually he strolls off, calling and listening, and eating.

Several hundred metres away is the distinctive shape of a group of turkeys waddling over the horizon. He'll finally have some buddies — possibly even female ones.

I've come full turkey circle.

I'm sustainable. Sort of.

Now, get growing, New Zealand.

No, really.

Get growing.